SO-DUE-669

PROBABILITY TABLES FOR LOCATING ELLIPTICAL
UNDERGROUND MASSES WITH A RECTANGULAR GRID

TABLITSY VEROYATNOSTEI PODSECHENIYA
ELLIPTICHESKIKH OB"EKTOV
PRYAMOUGOL'NOI SET'YU NABLYUDENII

ТАБЛИЦЫ ВЕРОЯТНОСТЕЙ ПОДСЕЧЕНИЯ
ЭЛЛИПТИЧЕСКИХ ОБЪЕКТОВ
ПРЯМОУГОЛЬНОЙ СЕТЬЮ НАБЛЮДЕНИЙ

Probability Tables
for Locating Elliptical Underground Masses with a Rectangular Grid

Igor' Dmitrievich Savinskii

Authorized translation from the Russian

CONSULTANTS BUREAU
NEW YORK
1965

The Russian text was published by Nedra in Moscow in 1964 for the State
Geological Committee of the USSR, All-Union Scientific-Research Institute
of Mineral Reserves (VIMS).

ИГОРЬ ДМИТРИЕВИЧ САВИНСКИЙ

ТАБЛИЦЫ ВЕРОЯТНОСТЕЙ ПОДСЕЧЕНИЯ ЭЛЛИПТИЧЕСКИХ ОБЪЕКТОВ
ПРЯМОУГОЛЬНОЙ СЕТЬЮ НАБЛЮДЕНИЙ

Library of Congress Catalog Card Number 65-20212

©1965 Consultants Bureau Enterprises, Inc.
227 West 17th St., New York, N.Y. 10011
All rights reserved

No part of this publication may be reproduced in any
form without written permission from the publisher

Printed in the United States of America

622.12
S267
t 1965

CONTENTS

PREFACE

The tables are intended for the determination of optimal rectangular observation grids and for the estimation of the reliability of prospecting regions when carrying out metallometric, borehole, magnetic, gravimetric, and other prospecting-exploratory work; they were computed for use by industrial workers in geophysics and geology. To make use of the tables it is first necessary to know the dimensions of the objects being sought; moreover, these objects must be assumed to be elliptical. As practical experience has shown, a number of geophysical, geochemical, and geological objects being sought can be very well represented by ellipses of various dimensions. In particular, objects being sought (for example, geophysical and geochemical anomalies) which occur in veins and in strata can be represented by ellipses with contraction coefficients of $0.1-0.2$, objects which are related to lenticular bodies, by ellipses with contraction coefficients of $0.3-0.5$, and objects which are related to isometric bodies (stockwork, ore shoots), by ellipses with contraction coefficients of $0.7-1.0$.

Existing computational methods are based on finding grids which will ensure that all the objects being sought will necessarily be detected; in this method it turns out to be sufficient to consider a certain unique position of the desired object relative to the observation grid. However, the grids obtained by such computations are unsuitable for use in many cases because of the high cost of the labor involved. It is customary, therefore, to use more open grids, but these create gaps which may miss objects which are of practical interest; as a rule, such grids cannot be justified mathematically. As a result, in practice grids are often used which are economically unprofitable, and work is frequently performed without an objective estimate of the reliability of regional prospecting. In recent years, during which mathematics has made deep inroads into the science of geology, the question of choosing and evaluating prospecting-exploratory grids has begun to be solved on the basis of theoretical computations of the probabilities of detection of the objects being sought under the condition that the objects are randomly distributed over the observation profiles. In the Soviet Union this trend was developed by A. P. Solovov and A. N. Eremeev, under whose initiative the computation of the present tables was formulated.

The tables presented here contain the values of the probability of interception of elliptic objects of various dimensions by different rectangular observation grids under the condition that these objects are randomly distributed. The computations were carried out separately for two cases: first, when to a certain extent the orientation of the object is known and the profiles are given perpendicular to the direction of elongation of the object with an error of $\pm30°$, and second, when the direction of the profiles has no relation to the orientation of the object.

For convenience, the tables are supplemented by a number of nomograms constructed from the data in the tables. These nomograms contain certain supplementary data; in particular, the unit probability isolines were derived by special computations. The probability of detection of the objects being sought, as found from the tables, leads to an objective characterization of the reliability of regional prospecting by given observation grids; this is of considerable value both for designing the grids for prospecting and for evaluating the reliability of regional prospecting carried out earlier by surveying. On the other hand, by use of the tables it is also possible to solve the converse problem — to find the optimal grid which guarantees a given detection probability for the objects being sought.

The author sincerely thanks A. N. Eremeev, V. M. Kreiter, I. Z. Samonov, A. P. Solovov, N. A. Khrushchov, and S. A. Shafranov for helpful discussions of the manuscript, and A. A. Korchagina for assistance in preparing the tables and nomograms for publication.

The author requests that his readers send remarks and questions to the following address: Moscow, V–17, Staromonetnyi per. 29, VIMS, Matematicheskaya Laboratoriya.

METHOD OF COMPUTING THE TABLES

The probability (P_n) of the interception of elliptical objects (Figs. 1, 3) by n points of a rectangular observation grid is computed on the basis of the assumption that with equal probability the ellipse may occupy.

(a) either any position in the xy-plane ($-\infty < x < +\infty$, $-\infty < y < +\infty$, $-90° \leq \Theta \leq +90°$; x, y are the coordinates of the center of the ellipse);

(b) or only the positions bounded by the angular interval $-30° \leq \Theta \leq +30°$ ($-\infty < x < +\infty$, $-\infty < y < +\infty$, $-30° \leq \Theta \leq +30°$).

The desired probability is defined as the ratio of the portion of the volume of the rectangular parallelepiped $h \times d/2 \times \pi/2$ for which the n points fall within the ellipse to the volume of the whole parallelepiped. The computations were carried out under the following constraints: (1) the major axis of the ellipse is less than the distance between the profiles ($a < d$); (2) the major axis of the ellipse is greater than the distance between the profiles but less than twice this distance ($d < a < 2d$). Different computation schemes were used in each of these cases.

Turning to the description of the calculation schemes, for the first case (see Fig. 1) we write the expression for $l(x, \Theta)$:

$$l(x, \Theta) = \frac{2a}{bc^2} \sqrt{\frac{b^2 c^2}{4} - x^2},$$

$$c^2 = \sin^2 \Theta + \frac{a^2}{b^2} \cos^2 \Theta.$$

For some fixed x and Θ if the ellipse intercepts a segment l of the profile and, moreover, if we allow freedom of movement in the y direction, the area of the ellipse will be intercepted by the $(n+1)$th point with the probability

$$P''_{n+1}(x, \Theta) = \frac{l}{h} - E\left(\frac{l}{h}\right),$$

where $E(l/h) = n$ is the integral part of l/h, and by n points with the probability $P''_n(x, \Theta) = 1 - P''_{n+1}(x, \Theta)$. In the case when simultaneously the

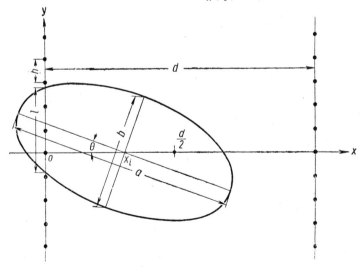

Fig. 1. Diagram for computing the probabilities in the case $a < d$.

3

ellipse has freedom of movement in the x direction, the probability $P'_{n+1}(\Theta)$ of its interception by the $(n+1)$th point will be

$$P'_{n+1}(\Theta) = \frac{2S_{n+1}}{dh},$$

where S_{n+1} is the area of the figure shown shaded in Fig. 2.

Under a displacement in the same interval the probability of interception by n points is

$$\frac{2}{dh}\left\{h \cdot x_n - \left[\int_0^{x_n} l(x,\Theta)\,dx - nhx_n\right]\right\}.$$

Consequently, for fixed Θ we get

$$P'_{n+1}(\Theta) = \frac{2}{dh}\left[\int_0^{x_n} l(x,\Theta)\,dx - nhx_n\right]$$

$$P'_n(\Theta) = \frac{2}{dh}\left\{h \cdot x_n - \left[\int_0^{x_n} l(x,\Theta)\,dx - nhx_n\right]\right\} +$$

$$+ \frac{2}{dh}\left[\int_{x_n}^{x_{n-1}} l(x,\Theta)\,dx - (n-1)\cdot h\cdot(x_{n-1}-x_n)\right].$$

$$\cdots\cdots\cdots\cdots\cdots\cdots$$

$$\cdots\cdots\cdots\cdots\cdots\cdots$$

$$P'_0(\Theta) = \frac{2}{dh}\left\{h\cdot(x_0-x_1) - \int_{x_1}^{x_0=bc} l(x,\Theta)\,dx\right\} +$$

$$+ \frac{d-bc}{d}.$$

Finally, the desired P_n would be found by numerical integration of $P'_n(\Theta)$ by Simpson's rule over the interval $(0, \pi/2)$.

In the second case $(d < a < 2d)$, when the ellipse may simultaneously intercept two profiles, the segment $(0, d/2)$ was subdivided by the points x_i into m equal intervals (Fig. 3). At each point x_i, l_1, l_2, and s were computed, and this allowed us to determine $P''_n(x_i, \Theta)$; next, $P'_n(\Theta)$ was found by summing the $P''_n(x_i, \Theta)$ over the segment $(0, d/2)$ by means of the trapezoid formula and by corresponding normalization.

In the determination of the $P''_n(x_i, \Theta)$ the integral parts $E(l_1/h)$, $E(l_2/h)$, and $E(s/h)$ were computed first, followed by the quantities $\Delta l_1 = (l_1/h) - E(l_1/h)$, $\Delta l_2 = (l_2/h) - E(l_1/h)$, and $\Delta S = (s/h) - E(s/h)$.

For fixed x and Θ and for freedom of displacement in the y direction the ellipse may be intercepted by $n = E(l_1/h) + E(l_2/h)$, $n+1$, and $n+2$ points. The corresponding probabilities, $P''_n(x, \Theta)$, $P''_{n+1}(x, \Theta)$, and $P''_{n+2}(x, \Theta)$ would be computed by starting with the following relations:

1. If $\Delta S + \Delta l_2 > 1$, then

when $\Delta l_1 < \Delta S + \Delta l_2 - 1$,
$$P''_{n+1} = 1 - \Delta l_2$$
$$P''_{n+1} = \Delta l_2 - \Delta l_1$$
$$P''_{n+2} = \Delta l_1$$

when $\Delta l_1 > \Delta S + \Delta l_2 - 1$, $\Delta l_1 > \Delta S$
$$P''_n = 0$$
$$P''_{n+1} = 2 - (\Delta l_1 + \Delta l_2)$$
$$P''_{n+2} = \Delta l_2 + \Delta l_1 - 1$$

when $\Delta l_1 < \Delta S + \Delta l_2$, $\Delta l_1 > \Delta S$
$$P''_n = 1 - (\Delta S + \Delta l_2)$$
$$P''_{n+1} = 2\Delta S + \Delta l_2 - \Delta l_1$$
$$P''_{n+2} = \Delta l_1 - \Delta S$$

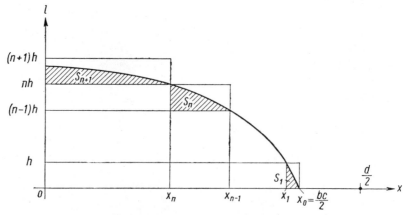

Fig. 2. Representation of the probabilities in the case $a < d$.

4

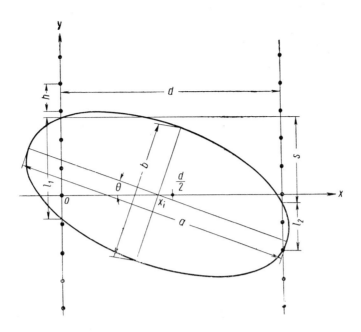

Fig. 3. Diagram for computing the probabilities in the case $d < a < 2d$.

2. If $\Delta S + \Delta l_2 < 1$, then

when $\Delta l_1 > \Delta S + \Delta l_2$
$$P_n'' = 1 - \Delta l_1$$
$$P''_{n+1} = \Delta l_1 - \Delta l_2$$
$$P''_{n+2} = \Delta l_2$$

when $\Delta l_1 < \Delta S + \Delta l_2, \quad \Delta l_1 < \Delta S$
$$P''_n = 1 + \Delta l_2 - \Delta l_1$$
$$P''_{n+1} = \Delta l_1 + \Delta l_2$$
$$P''_{n+2} = 0$$

when $\Delta l_1 > \Delta S + \Delta l_2 - 1, \quad \Delta l_1 < \Delta S$
$$P''_n = \Delta S - \Delta l_1$$
$$P''_{n+1} = \Delta l_1 - \Delta l_2 - 2\Delta S$$
$$P''_{n+2} = \Delta l_2 + \Delta S - 1$$

In the actual computations the segment $(0, \pi/2)$ was subdivided into 30–90 equal intervals and the segment $(0, d/2)$ into 50–500 intervals; this ensured the number of significant digits shown in the tables. The computations were carried out on a Strela-3 electronic computer at the Computation Center of the Academy of Sciences of the USSR.

5

HOW TO USE THE TABLES

Each table consists of the values of the probabilities computed for the indicated parameters: the contraction coefficient b', the distance d between the profiles (expressed in units of the length a of the object), and the range of Θ ($-90° \leq \Theta \leq +90°$ or $-30° \leq \Theta \leq +30°$). Six values of b' were selected: 1.0, 0.7, 0.5, 0.3, 0.2, and 0.1. In each case the distance d between the profiles was varied from 0.5 to 2.0.

The range $-30° \leq \Theta \leq +30°$ corresponds to the case when the profiles of the prospecting grid are taken perpendicular to the direction of elongation of the objects being sought with a maximum error of $\pm 30°$ (for a uniform distribution of error in this range); the range $-90° \leq \Theta \leq +90°$ corresponds to the case when the direction of the profile is unrelated to the orientation of the objects being sought.

The columns of the tables consist of the values of the probabilities, computed for various values of h (the profile spacing), expressed in units of d. The magnitude of h varies from $h = 0.05d$ to $h = d$ (a square grid). In the columns P_1, P_2, ..., P_7 are located the probabilities of the interception of the object, respectively, by one, two, ..., seven points of the grid. In the column P_0 are located the probabilities of not intercepting the object. In the column $P_{\geq 8}$ are located the probabilities of intercepting of the object by eight or more points of the grid.

By summing some of the indicated probabilities we can compute the probability of intercepting the object by one or more points of the grid $P_{\geq 1} = P_1 + P_2 + \ldots + P_7 + P_{\geq 8}$, and the probability of interception the object by two or more points of the grid, $P_{\geq 2} = P_2 + P_3 + \ldots + P_7 + P_{\geq 8}$. Similarly, we can compute the probability of intercepting the object by three or more points, $P_{\geq 3} = P_3 + P_4 + \ldots + P_7 + P_{\geq 8}$, etc. The last column of the tables gives the probabilities of intercepting the object simultaneously by two observation profiles (without regard to the number of intercepted points).

The bottom row of the tables gives the probability of intercepting the object in the case of continuous observation along the profile. In this row the P_0 column indicates the probability of noninterception of the object by any of the profiles, and the last column the probability of its interception simultaneously by two profiles. The probability of intercepting the object by one profile can be computed by subtracting the sum of these indicated probabilities from unity. Thus, for example, when $b' = 0.2$, $-90° \leq \Theta \leq +90°$, $d = 0.5$ (Table 98), the probability of noninterception of the object by a profile of continuous observations is 0.11; the probability of simultaneous interception by two of them is 0.45; the probability of interception by one profile is $1 - (0.11 + 0.45) = 0.44$.

Let us consider the example of using the tables to compute the probability of detecting the objects being sought when working with certain given observation grids. The computations should be based on certain assumptions concerning the dimensions of the objects being sought; we shall not consider the question of the determination of these dimensions even though it has independent significant importance. It is essential to emphasize that in all the subsequent computations we assume that the positions of the objects being sought relative to each other are independent.

Suppose that it is known that the objects being sought are anomalies 50% of which have the dimensions $a_1 = 400$ m, $b_1 = 200$ m ($b_1' = 0.5$), 30% the dimensions $a_2 = 500$ m, $b_2 = 100$ m ($b_2' = 0.2$), and 20% the dimensions $a_3 = 600$ m, $b_3 = 200$ m ($b_3' = 0.3$). Thus we have three different types of anomalies. Moreover, it is known that there exist specific orientations of the anomalies and that the profiles of the prospecting grid are perpendicular to the direction of elongation of the anomalies with an

6

error of $\pm 30°$. We wish to determine the probability $P_{\geqslant 1}$ of detecting the anomalies when prospecting with a grid of $d = 500$ m and $h = 100$ m.

Let us express the distance d in units of a_1 and the distance h in units of d; then $d = 1.25a_1$, $h = 0.2d$. In what follows, instead of the value $d = 1.25$, for which there is no table, we shall take the value $d = 1.3$, as a result of which the computed detection probability will be somewhat lower (we could also have had recourse to interpolation). By assuming $b' = 0.5$, $d = 1.3$, $h = 0.2$ from Table 61, we find the probability of detecting anomalies of the first type: $P_{\geqslant 1}^{\mathrm{I}} = 0.711$. Similarly, expressing d in units of a_2 ($d = 1.00a_2$) and taking $b' = 0.2$, $d = 1.0$, $h = 0.2$, from Table 107 we find the probability of detecting anomalies of the second type: $P_{\geqslant 1}^{\mathrm{II}} = 0.774$. For anomalies of the third type we have $b' = 0.3$, $d = 0.9$, $h = 0.2$; from Table 79 we find $P_{\geqslant 1}^{\mathrm{III}} = 0.98$. We can now compute the average probability of detecting the anomalies (taking the percentage occurrences of the different anomaly types into account):

$$\overline{P} = 0.5 \cdot 0.711 + 0.3 \cdot 0.774 + 0.2 \cdot 0.98 = 0.78.$$

Thus, under the stated conditions, on the average 78 of 100 anomalies will be detected and 22 will be missed.

Under actual conditions the anomalies often have a tendency to occur in groups. The problem of detecting at least one anomaly in each different group is therefore of practical interest. The probability P^* of detecting groups of anomalies in the first approximation (by assuming independence of the locations of the anomalies relative to each other) can be computed from the formula

$$P^* = 1 - (1 - P^{\mathrm{I}})(1 - P^{\mathrm{II}}) \ldots (1 - P^r),$$

where P^{I}, P^{II}, ..., P^r are the probabilities found from the tables of detecting each of the r anomalies occurring in the group.

If all the anomalies occurring in the group have the same dimensions, the formula takes the form:[†]

$$\overline{P}^* = 1 - (1 - P)^r.$$

Consider the following example. Let each group consist of three anomalies having the same dimensions $a = 100$ m, $b = 50$ m ($b' = 0.5$); the survey is being carried out by means of a grid with $d = 150$ m and $h = 15$ m ($h = 0.1d$). We wish to compute the probability P of detecting the groups under the con-

dition that the orientation of the profile has no relation to the orientation of the anomalies ($-90° \leqslant \Theta \leqslant +90°$) and that for the detection of each anomaly it is sufficient to intercept it by even one point. Taking $b' = 0.5$, $d = 1.5$, and $h = 0.1$, from Table 66 we find that $P_{\geqslant 1} = 0.509$. Consequently, $P^* = 1 - (1 - 0.509)^3 = 0.882$.

Let us now consider the question of choosing the optimal observation grid. Most simply this problem is solved in the case when the number N of observation points per square kilometer is given a priori and it is required only to find the optimal relation between d and h. If the objects being sought have different dimensions, we may compute the average probability $\overline{P}$ by the method indicated above for all the practically admissible grids corresponding to the given N, and then choose from these grids that one for which the average probability turns out to be the highest (method of selection).

In the more general case the problem consists in choosing the value of the probability with which the region should be prospected and finding the economically most advantageous grid which will ensure this probability. To solve this problem some economic index which characterizes the grid must be taken into consideration. In the examples that follow we shall everywhere assume (as a first approximation) that the cost of labor is directly proportional to the number of observation points. Thus, the average number of points necessary for the observation of each anomaly may be used as the economic index. Denoting this number by K^*, we have

$$K^* = \frac{N}{P \cdot M},$$

when N is the number of grid points per square kilometer, P is the probability of detecting an anomaly (in particular cases, for example, $P_{\geqslant 1}$ or $P_{\geqslant 2}$), and M is the number of anomalies per square kilometer in the given region.

However, in order to have an index which characterizes the same grid independently of the quantity M, it is convenient to use the coefficient $K = N/P$, numerically equal to K^* when $M = 1$. A comparison of coefficients K for different conditions in one and the same region (or for different regions having the same M) is equivalent to a comparison of the corresponding K^*. Such a comparison allows us to compare the expenditures of detecting every

---

[†]The method of computation cited here has been previously used by A.P. Solovov.

7

individual object under various conditions of prospecting by eliminating the quantity M from consideration.

If d and h are expressed in meters, $K = 10^6/dhP$; if d and h are expressed in units of a, but a is expressed in meters, $K = 10^6/dhPa^2$. In the example considered above $K = 10^6/(500 \cdot 100 \cdot 0.78) = 26$, i.e., if in the region there is on the average one anomaly per square kilometer, 26 points per observation will be required (if there are ten anomalies per square kilometer, 2.6 points, etc.). If in the cited example the profiles are taken without relation to the direction of elongation of the anomalies, and interception by not less than two grid points is considered necessary for detecting the anomalies, we obtain $P_{\geqslant 2}^{I} = 0.432$, $P_{\geqslant 2}^{II} = 0.140$, and $P_{\geqslant 2}^{III} = 0.49$ (by using the data in Tables 62, 108, 80). Here

$$K = \frac{10^6}{500 \cdot 100 \cdot 0.36} = 56.$$

Thus, in this case the expenditure per detected anomaly turns out to be approximately twice as large as before.

Let us now consider an example of finding the most open grid (the one having the least number of points per square kilometer) which would ensure a given detection probability. Let the objects being sought have the dimensions $a = 100$ m, $b = 20$ m ($b' = 0.2$), and let us pose the problem of prospecting the region with a detection probability of $P_{\geqslant 2} = 0.8$ (i.e., missing on the average 20 out of 100 objects can be tolerated); we assume that the orientation of the profile has no relation to the orientation of the objects being sought ($-90° \leq \Theta \leq +90°$). The problem is conveniently solved by using the nomograms.

The nomograms are intended for choosing the best grid in the case when the objects being sought have the same dimensions and only the probabilities $P_{\geqslant 1}$ or $P_{\geqslant 2}$ are considered; in all other cases one has to use the tables directly and employ the method of selection. The construction of each nomogram starts with the nature of the probability of the event (only $P_{\geqslant 1}$ or $P_{\geqslant 2}$ is considered), the range of Θ, and the contraction coefficient b'. Along the ordinate axis are plotted the values of dh and of their logarithms; d and h are expressed in units of a. Along the abscissa axis is plotted h, the profile spacing, expressed in units of d. In this coordinate system curves are plotted for d from $d = 0.5$ to $d = 0.2$ (d is expressed in units of

a). Every point on the d curves corresponds to a specific observation grid (obtainable by reading off the corresponding abscissa h) — the greater the ordinate of the point the more open the grid. The d curves are intersected by isolines of probabilities 0.1, 0.2, 0.3, ..., 1.0; the point of intersection of any isoline with a d curve indicates that this is the observation grid which ensures the detection probability corresponding to the given isoline.

For the example being considered we choose Nomogram 10, which has the designations $P_{\geqslant 2}$, $-90° \leq \Theta \leq +90°$, $b' = 0.2$. On this nomogram we locate the isoline of probability 0.8 and from among the points of intersection of this isoline with the d curves we choose that point which has the maximum ordinate. This point is the point of intersection of the isoline with the curve for $d = 0.5$. The grid corresponding to this point of intersection has the dimensions $d = 0.5a$, $h = 0.28a$. In our example $a = 100$ m, whence we get $d = 50$ m, $h = 0.28 \cdot 50$ m $= 14$ m. This is the most open grid (of all those derived in the tables) which ensures a detection probability of $P_{\geqslant 2} = 0.8$ under the prospecting conditions cited in the example.

On the nomograms the values of $K' = 1/dhP$ (d and h expressed in units of a) are for grids which correspond to points of intersection of isolines having the largest ordinates. For the grid we found above this value equals 18. In the case when the length a of the object remains constant a comparison of the values of K' for different conditions of prospecting is equivalent to a comparison of the coefficients $K = 10^6/dhPa^2$ considered above. Since in setting up the nomograms it was assumed that $a = 1$ everywhere, a comparison of the values of K' indicated on them allows us to obtain a graphic representation of the relative expenditures for the detection of every individual object under various given detection probabilities, various reliabilities of pin-pointing the anomalies (only $P_{\geqslant 1}$ or $P_{\geqslant 2}$ is considered), various ranges of Θ, and various contractions b'.

The choice of the detection probability with which we should explore a region is, in the general case, necessarily made with due regard to the indicated values of K'. As can be seen from the nomograms, as the probability $P_{\geqslant 2}$ increases (see the corresponding isolines) the values of K' at first decrease but then, having reached a minimum, again rise. Obviously it is not reasonable to work with probabilities for which K' still has not reached a minimum (probabilities of 0.1–0.6 in the example being considered), since in that case there occur both a significant overlooking of the objects being sought as well as an excessive expenditure for the detection of each object. When there is an increase in the values of the probabilities for which the considered value of K' corresponds to the minimum or passes through it (probabilities of 0.7−0.85), the reliability of prospecting the region also will increase, but on the other hand there will be an increase in the expense of detecting the objects as well. It is obvious that in this case (and also when considering the probability $P_{\geqslant 1}$) we should start with some compromise solution based on an actual accounting of complex economic, geographic, and other factors.

When the problem is posed of detecting objects of different dimensions, the choice of the most reasonable grid can be made in the following way. First, the average probability $\overline{P}$ and the corresponding values of K for all the grids which can be used in practice are found by the same method as in the example considered above. Next, the grids with the same (or nearly same) probability $\overline{P}$ are grouped, and from each group the grid with the least value of K is chosen. From the grids so selected we then eliminated all grids having a lower $\overline{P}$ and a larger K than other grids. The remaining grids may be used; the choice among them should be made on basis of comparing the $\overline{P}$ and K values of each grid, and reaching a compromise solution.

In the previous examples the most open grid was taken to be the most advantageous. However, in many practical cases the expense of prospecting will be determined not only by the number of grid points per square kilometer but also by the relation between d and h; with decreasing distance between points along the profile and a simultaneous increase in the distance between profiles, the labor cost will, as a rule, decrease. In the computations performed above the corresponding corrections can be easily inserted if instead of the quantity N (the number of points per square kilometer) we everywhere substitute the quantity W (the cost of carrying out the work per square kilometer); here, instead of the coefficient $K = N/P$ we take the coefficient $t = W/P$ to characterize the cost of detecting one anomaly. In particular, for every actual prospecting method we can construct a nomogram analogous to those in this book, in which instead of the quantity dh along the ordinate axis, we plot the quantity $1/W$. Then, by choosing from among the points of intersection of a given probability isoline with the d curves (which, obviously, will be shaped differently) the point with the maximum ordinate, we can obtain not the most open but the cheapest prospecting grid.

9

TABLES OF THE PROBABILITY OF
INTERCEPTION OF ELLIPTIC OBJECTS
BY A RECTANGULAR OBSERVATION GRID

Table 1

Circle   x spacing / diameter

$$b' = 1{,}0 \qquad d = 0{,}50$$

h in units of d	P_0	P_1	P_2	P_3	P_4	P_5	P_6	P_7	$P_{\geqslant 8}$	$P_{\geqslant 1}$	$P_{\geqslant 2}$	Probability of interception by two profiles
0,05									1,000	1,000	1,000	0,999
0,10									1,000	1,000	1,000	0,999
0,15									1,000	1,000	1,000	0,998
0,20									1,000	1,000	1,000	0,996
0,25								<0,0005	1,000	1,000	1,000	0,995
0,30							0,005	0,007	0,987	1,000	1,000	0,992
0,35						<0,0005	0,041	0,043	0,916	1,000	1,000	0,990
0,40					<0,0005	0,014	0,095	0,168	0,723	1,000	1,000	0,986
0,45					0,016	0,034	0,275	0,305	0,370	1,000	1,000	0,983
0,50				<0,0005	0,024	0,129	0,467	0,300	0,080	1,000	1,000	0,979
0,60				0,009	0,202	0,333	0,456			1,000	1,000	0,969
0,70			0,016	0,032	0,513	0,329	0,111			1,000	1,000	0,958
0,80			0,048	0,103	0,740	0,091	0,017			1,000	1,000	0,945
0,90			0,074	0,362	0,564					1,000	1,000	0,930
1,00		<0,0005	0,174	0,511	0,315					1,000	1,000	0,913
Continuous observation along the profile	—											1,00

Square → (annotation beside row 1,00)

Gilbert 1-β   10.3   g=1,0   L/G=1,0

Table 2

$$b' = 1{,}0 \qquad d = 0{,}55$$

h in units of d	P_0	P_1	P_2	P_3	P_4	P_5	P_6	P_7	$P_{\geqslant 8}$	$P_{\geqslant 1}$	$P_{\geqslant 2}$	Probability of interception by two profiles
0,05									1,000	1,000	1,000	0,818
0,10									1,000	1,000	1,000	0,817
0,15									1,000	1,000	1,000	0,816
0,20									1,000	1,000	1,000	0,815
0,25								0,137	0,863	1,000	1,000	0,812
0,30							0,181	0,058	0,761	1,000	1,000	0,810
0,35						0,155	0,106	0,126	0,614	1,000	1,000	0,807
0,40					0,097	0,122	0,243	0,269	0,269	1,000	1,000	0,803
0,45					0,199	0,119	0,438	0,201	0,043	1,000	1,000	0,799
0,50				0,068	0,232	0,147	0,543	0,009	0,001	1,000	1,000	0,795
0,60			<0,0005	0,210	0,358	0,326	0,106			1,000	1,000	0,785
0,70			0,111	0,179	0,614	0,082	0,014			1,000	1,000	0,772
0,80			0,199	0,356	0,445					1,000	1,000	0,758
0,90			0,346	0,422	0,231					1,000	1,000	0,741
1,00		0,034	0,488	0,325	0,153					1,000	0,966	0,722
Continuous observation along the profile	—											0,818

13

Table 3

$$b'=1,0 \qquad d=0,60$$

h in units of d	P_0	P_1	P_2	P_3	P_4	P_5	P_6	P_7	$P_{\geqslant 8}$	$P_{\geqslant 1}$	$P_{\geqslant 2}$	Probability of interception by two profiles
0,05									1,000	1,000	1,000	0,666
0,10									1,000	1,000	1,000	0,665
0,15									1,000	1,000	1,000	0,664
0,20									1,000	1,000	1,000	0,663
0,25							0,131	0,217	0,651	1,000	1,000	0,660
0,30						0,162	0,226	0,062	0,550	1,000	1,000	0,658
0,35					0,099	0,257	0,192	0,218	0,235	1,000	1,000	0,654
0,40					0,305	0,140	0,378	0,150	0,027	1,000	1,000	0,650
0,45				0,108	0,332	0,164	0,396			1,000	1,000	0,646
0,50				0,240	0,314	0,288	0,158			1,000	1,000	0,641
0,60			0,104	0,292	0,486	0,100	0,018			1,000	1,000	0,630
0,70			0,265	0,354	0,381					1,000	1,000	0,616
0,80			0,452	0,369	0,179					1,000	1,000	0,600
0,90		0,054	0,576	0,261	0,109					1,000	0,946	0,582
1,00		0,120	0,639	0,180	0,061					1,000	0,880	0,570
Continuous observation along the profile	—											0,667

Table 4

$$b'=1,0 \qquad d=0,65$$

h in units of d	P_0	P_1	P_2	P_3	P_4	P_5	P_6	P_7	$P_{\geqslant 8}$	$P_{\geqslant 1}$	$P_{\geqslant 2}$	Probability of interception by two profiles
0,05									1,000	1,000	1,000	0,537
0,10									1,000	1,000	1,000	0,537
0,15									1,000	1,000	1,000	0,536
0,20								0,197	0,803	1,000	1,000	0,534
0,25						0,007	0,439	0,069	0,484	1,000	1,000	0,532
0,30					0,009	0,434	0,152	0,161	0,244	1,000	1,000	0,529
0,35					0,323	0,217	0,308	0,130	0,022	1,000	1,000	0,525
0,40				0,099	0,452	0,153	0,296			1,000	1,000	0,521
0,45				0,302	0,366	0,232	0,101			1,000	1,000	0,516
0,50			0,010	0,464	0,353	0,142	0,030			1,000	1,000	0,511
0,60			0,257	0,388	0,355					1,000	1,000	0,499
0,70			0,494	0,356	0,150					1,000	1,000	0,484
0,80		0,049	0,658	0,213	0,080					1,000	0,951	0,466
0,90		0,151	0,673	0,136	0,040					1,000	0,849	0,446
1,00		0,242	0,672	0,071	0,015					1,000	0,758	0,422
Continuous observation along the profile	—											0,538

Table 5

$$b' = 1{,}0 \qquad d = 0{,}70$$

h in units of d	P_0	P_1	P_2	P_3	P_4	P_5	P_6	P_7	$P_{\geqslant 8}$	$P_{\geqslant 1}$	$P_{\geqslant 2}$	Probability of interception by two profiles
0,05									1,000	1,000	1,000	0,428
0,10									1,000	1,000	1,000	0,428
0,15									1,000	1,000	1,000	0,426
0,20							0,061	0,495	0,444	1,000	1,000	0,424
0,25						0,256	0,373	0,090	0,281	1,000	1,000	0,421
0,30					0,221	0,395	0,233	0,125	0,027	1,000	1,000	0,418
0,35				0,033	0,590	0,142	0,235			1,000	1,000	0,414
0,40				0,308	0,444	0,181	0,068			1,000	1,000	0,410
0,45			0,012	0,533	0,351	0,089	0,015			1,000	1,000	0,405
0,50			0,150	0,503	0,340	0,007	<0,0005			1,000	1,000	0,399
0,60			0,465	0,398	0,137					1,000	1,000	0,385
0,70		0,016	0,739	0,183	0,062					1,000	0,984	0,369
0,80		0,154	0,715	0,104	0,027					1,000	0,846	0,350
0,90		0,279	0,669	0,044	0,008					1,000	0,721	0,327
1,00		0,401	0,595	0,003	<0,0005					1,000	0,599	0,301
Continuous observation along the profile	—											0,429

Table 6

$$b' = 1{,}0 \qquad d = 0{,}75$$

h in units of d	P_0	P_1	P_2	P_3	P_4	P_5	P_6	P_7	$P_{\geqslant 8}$	$P_{\geqslant 1}$	$P_{\geqslant 2}$	Probability of interception by two profiles
0,05									1,000	1,000	1,000	0,333
0,10									1,000	1,000	1,000	0,332
0,15								0,013	0,988	1,000	1,000	0,331
0,20						0,009	0,417	0,288	0,286	1,000	1,000	0,328
0,25					0,042	0,539	0,257	0,117	0,045	1,000	1,000	0,326
0,30				0,006	0,530	0,267	0,197			1,000	1,000	0,322
0,35				0,243	0,572	0,138	0,047			1,000	1,000	0,318
0,40			0,012	0,550	0,379	0,051	0,007			1,000	1,000	0,313
0,45			0,132	0,633	0,235					1,000	1,000	0,308
0,50			0,344	0,520	0,136					1,000	1,000	0,301
0,60		0,003	0,716	0,232	0,049					1,000	0,997	0,287
0,70		0,121	0,781	0,079	0,019					1,000	0,879	0,269
0,80		0,287	0,683	0,026	0,004					1,000	0,713	0,248
0,90	<0,0005	0,448	0,552							1,000	0,552	0,224
1,00	0,007	0,590	0,403							0,993	0,403	0,201
Continuous observation along the profile	—											0,333

Table 7

$$b'=1,0 \qquad d=0,80$$

h in units of d	P_0	P_1	P_2	P_3	P_4	P_5	P_6	P_7	$P_{\geqslant 8}$	$P_{\geqslant 1}$	$P_{\geqslant 2}$	Probability of interception by two profiles
0,05									1,000	1,000	1,000	0,250
0,10									1,000	1,000	1,000	0,249
0,15							0,014	0,207	0,779	1,000	1,000	0,247
0,20					<0,0005	0,175	0,592	0,153	0,079	1,000	1,000	0,245
0,25					0,260	0,570	0,169	<0,0005		1,000	1,000	0,242
0,30				0,117	0,708	0,141	0,033			1,000	1,000	0,238
0,35			0,012	0,499	0,461	0,025	0,003			1,000	1,000	0,233
0,40			0,107	0,719	0,174					1,000	1,000	0,228
0,45			0,348	0,578	0,075					1,000	1,000	0,222
0,50			0,585	0,375	0,040					1,000	1,000	0,216
0,60		0,059	0,850	0,078	0,013					1,000	0,941	0,200
0,70		0,262	0,724	0,012	0,001					1,000	0,738	0,181
0,80	0,001	0,464	0,535							0,999	0,535	0,160
0,90	0,010	0,616	0,374							0,990	0,374	0,142
1,00	0,028	0,716	0,255							0,972	0,255	0,128
Continuous observation along the profile	—											0,250

Table 8

$$b'=1,0 \qquad d=0,85$$

h in units of d	P_0	P_1	P_2	P_3	P_4	P_5	P_6	P_7	$P_{\geqslant 8}$	$P_{\geqslant 1}$	$P_{\geqslant 2}$	Probability of interception by two profiles
0,05									1,000	1,000	1,000	0,176
0,10									1,000	1,000	1,000	0,175
0,15						0,013	0,174	0,380	0,433	1,000	1,000	0,173
0,20					0,083	0,410	0,497	0,009	0,001	1,000	1,000	0,171
0,25				0,063	0,548	0,367	0,022			1,000	1,000	0,168
0,30			0,014	0,357	0,623	0,006	<0,0005			1,000	1,000	0,164
0,35			0,101	0,692	0,207					1,000	1,000	0,159
0,40			0,316	0,650	0,034					1,000	1,000	0,153
0,45		0,004	0,593	0,385	0,018					1,000	0,996	0,147
0,50		0,031	0,771	0,189	0,008					1,000	0,969	0,140
0,60		0,192	0,805	0,003	<0,0005					1,000	0,808	0,123
0,70	0,003	0,441	0,556							0,997	0,556	0,106
0,80	0,014	0,612	0,373							0,986	0,373	0,093
0,90	0,034	0,725	0,242							0,966	0,242	0,082
1,00	0,061	0,791	0,148							0,939	0,148	0,074
Continuous observation along the profile	—											0,176

16

Table 9

b′ = 1,0   d = 0,90

h in units of d	P_0	P_1	P_2	P_3	P_4	P_5	P_6	P_7	$P_{\geqslant 8}$	$P_{\geqslant 1}$	$P_{\geqslant 2}$	Probability of interception by two profiles
0,05									1,000	1,000	1,000	0,110
0,10							0,005	0,068	0,927	1,000	1,000	0,109
0,15					0,022	0,140	0,293	0,445	0,101	1,000	1,000	0,108
0,20				0,039	0,265	0,504	0,192			1,000	1,000	0,105
0,25			0,018	0,230	0,607	0,145				1,000	1,000	0,102
0,30			0,108	0,551	0,340					1,000	1,000	0,097
0,35		0,001	0,281	0,664	0,054					1,000	0,999	0,092
0,40		0,020	0,541	0,436	0,004					1,000	0,980	0,087
0,45		0,061	0,724	0,214	0,001					1,000	0,939	0,080
0,50	<0,0005	0,133	0,794	0,073						1,000	0,867	0,073
0,60	0,006	0,372	0,622							0,994	0,622	0,060
0,70	0,019	0,576	0,404							0,981	0,404	0,052
0,80	0,040	0,709	0,252							0,961	0,252	0,045
0,90	0,067	0,789	0,144							0,933	0,144	0,040
1,00	0,103	0,825	0,073							0,897	0,073	0,036
Continuous observation along the profile	—											0,111

Table 10

b′ = 1,0   d = 0,95

h in units of d	P_0	P_1	P_2	P_3	P_4	P_5	P_6	P_7	$P_{\geqslant 8}$	$P_{\geqslant 1}$	$P_{\geqslant 2}$	Probability of interception by two profiles
0,05									1,000	1,000	1,000	0,052
0,10					0,005	0,050	0,104	0,102	0,739	1,000	1,000	0,051
0,15			<0,0005	0,040	0,145	0,164	0,276	0,374	0,001	1,000	1,000	0,049
0,20			0,038	0,164	0,264	0,476	0,058			1,000	1,000	0,046
0,25		0,002	0,135	0,290	0,527	0,046				1,000	0,998	0,043
0,30		0,020	0,247	0,544	0,188					1,000	0,980	0,038
0,35	<0,0005	0,062	0,390	0,548	<0,0005					1,000	0,938	0,033
0,40	0,002	0,116	0,586	0,296						0,998	0,882	0,029
0,45	0,006	0,179	0,689	0,125						0,994	0,815	0,026
0,50	0,012	0,259	0,706	0,023						0,988	0,729	0,023
0,60	0,027	0,496	0,477							0,973	0,477	0,019
0,70	0,048	0,661	0,291							0,952	0,291	0,017
0,80	0,075	0,763	0,162							0,925	0,162	0,015
0,90	0,109	0,816	0,076							0,891	0,076	0,013
1,00	0,153	0,824	0,023							0,847	0,023	0,012
Continuous observation along the profile	—											0,053

Table 11

b′=1,0   d=1,0

h in units of d	P_0	P_1	P_2	P_3	P_4	P_5	P_6	P_7	$P_{\geqslant 8}$	$P_{\geqslant 1}$	$P_{\geqslant 2}$	Probability of interception by two profiles
0,05	0,001	0,003	0,005	0,008	0,010	0,013	0,016	0,019	0,927	0,999	0,997	
0,10	0,002	0,010	0,020	0,032	0,044	0,058	0,075	0,099	0,660	0,998	0,988	
0,15	0,004	0,023	0,048	0,076	0,114	0,176	0,364	0,195		0,996	0,973	
0,20	0,007	0,041	0,089	0,154	0,301	0,408				0,993	0,952	
0,25	0,011	0,066	0,149	0,322	0,453					0,989	0,924	
0,30	0,015	0,097	0,240	0,550	0,098					0,985	0,888	
0,35	0,021	0,137	0,421	0,422						0,979	0,843	
0,40	0,027	0,186	0,582	0,204						0,973	0,786	
0,45	0,035	0,250	0,650	0,065						0,965	0,715	
0,50	0,043	0,343	0,614							0,957	0,614	
0,60	0,064	0,564	0,373							0,936	0,373	
0,70	0,089	0,700	0,211							0,911	0,211	
0,80	0,121	0,777	0,102							0,879	0,102	
0,90	0,160	0,807	0,033							0,840	0,033	
1,00	0,215	0,785								0,785		
Continuous observation along the profile	—											

Table 12

b′=1,0   d=1,1

h in units of d	P_0	P_1	P_2	P_3	P_4	P_5	P_6	P_7	$P_{\geqslant 8}$	$P_{\geqslant 1}$	$P_{\geqslant 2}$	Probability of interception by two profiles
0,05	0,091	0,003	0,006	0,008	0,011	0,014	0,017	0,021	0,828	0,909	0,906	
0,10	0,093	0,011	0,023	0,035	0,049	0,066	0,089	0,123	0,511	0,907	0,896	
0,15	0,095	0,025	0,053	0,087	0,135	0,241	0,360	0,005		0,905	0,880	
0,20	0,098	0,046	0,100	0,183	0,414	0,159				0,902	0,856	
0,25	0,103	0,073	0,172	0,430	0,222					0,898	0,824	
0,30	0,108	0,109	0,298	0,483	0,003					0,892	0,784	
0,35	0,114	0,155	0,495	0,237						0,886	0,732	
0,40	0,121	0,214	0,585	0,080						0,879	0,664	
0,45	0,130	0,300	0,568	0,002						0,870	0,570	
0,50	0,139	0,424	0,437							0,861	0,437	
0,60	0,162	0,594	0,244							0,838	0,244	
0,70	0,191	0,690	0,118							0,809	0,118	
0,80	0,228	0,732	0,040							0,772	0,040	
0,90	0,280	0,719	0,001							0,720	0,001	
1,00	0,351	0,649								0,649		
Continuous observation along the profile	0,091											

Table 13

$b' = 1,0$ $d = 1,2$

h in units of d	P_0	P_1	P_2	P_3	P_4	P_5	P_6	P_7	$P_{\geqslant 8}$	$P_{>1}$	$P_{\geqslant 2}$	Probability of interception by two profiles
0,05	0,167	0,003	0,006	0,009	0,012	0,016	0,019	0,023	0,744	0,833	0,830	
0,10	0,169	0,012	0,025	0,039	0,055	0,076	0,105	0,163	0,357	0,832	0,819	
0,15	0,171	0,028	0,059	0,098	0,162	0,347	0,136			0,829	0,801	
0,20	0,175	0,050	0,112	0,225	0,412	0,026				0,825	0,775	
0,25	0,179	0,081	0,200	0,458	0,082					0,821	0,740	
0,30	0,185	0,121	0,384	0,309						0,815	0,694	
0,35	0,192	0,175	0,517	0,117						0,808	0,633	
0,40	0,200	0,250	0,537	0,013						0,800	0,550	
0,45	0,209	0,370	0,421							0,791	0,421	
0,50	0,220	0,470	0,311							0,780	0,311	
0,60	0,246	0,600	0,155							0,754	0,155	
0,70	0,279	0,662	0,058							0,721	0,058	
0,80	0,325	0,669	0,007							0,675	0,007	
0,90	0,394	0,606								0,606		
1,00	0,455	0,545								0,545		
Continuous observation along the profile	0,167											

Table 14

$b' = 1,0$ $d = 1,3$

h in units of d	P_0	P_1	P_2	P_3	P_4	P_5	P_6	P_7	$P_{\geqslant 8}$	$P_{>1}$	$P_{\geqslant 2}$	Probability of interception by two profiles
0,05	0,231	0,003	0,007	0,010	0,013	0,017	0,021	0,026	0,672	0,769	0,766	
0,10	0,233	0,013	0,027	0,043	0,061	0,087	0,128	0,259	0,149	0,767	0,754	
0,15	0,236	0,030	0,064	0,111	0,203	0,342	0,015			0,764	0,734	
0,20	0,240	0,055	0,126	0,301	0,278					0,760	0,706	
0,25	0,245	0,089	0,238	0,419	0,009					0,756	0,667	
0,30	0,251	0,135	0,429	0,185						0,749	0,614	
0,35	0,258	0,198	0,501	0,042						0,742	0,544	
040	0,267	0,304	0,429							0,733	0,429	
0,45	0,277	0,413	0,310							0,723	0,310	
0,50	0,289	0,492	0,218							0,711	0,218	
0,60	0,318	0,589	0,093							0,682	0,093	
0,70	0,357	0,621	0,021							0,643	0,021	
0,80	0,419	0,581								0,581		
0,90	0,484	0,516								0,516		
1,00	0,535	0,465								0,465		
Continuous observation along the profile	0,231											

Table 15

b′=1,0   d=1,4

h in units of d	P_0	P_1	P_2	P_3	P_4	P_5	P_6	P_7	$P_{\geqslant 8}$	$P_{\geqslant 1}$	$P_{\geqslant 2}$	Probability of interception by two profiles
0,05	0,286	0,004	0,007	0,011	0,015	0,019	0,023	0,028	0,608	0,714	0,710	
0,10	0,288	0,014	0,029	0,047	0,068	0,100	0,167	0,273	0,014	0,712	0,698	
0,15	0,291	0,033	0,071	0,126	0,280	0,200				0,709	0,676	
0,20	0,295	0,060	0,142	0,353	0,150					0,705	0,645	
0,25	0,301	0,098	0,300	0,302						0,699	0,602	
0,30	0,307	0,150	0,443	0,100						0,693	0,543	
0,35	0,315	0,228	0,453	0,004						0,685	0,457	
0,40	0,325	0,348	0,327							0,675	0,327	
0,45	0,336	0,437	0,227							0,664	0,227	
0,50	0,349	0,500	0,151							0,651	0,151	
0,60	0,382	0,568	0,050							0,618	0,050	
0,70	0,429	0,569	0,002							0,571	0,002	
0,80	0,499	0,501								0,501		
0,90	0,555	0,445								0,445		
1,00	0,599	0,401								0,401		
Continuous observation along the profile	0,286											

Table 16

b′=1,0   d=1,5

h in units of d	P_0	P_1	P_2	P_3	P_4	P_5	P_6	P_7	$P_{\geqslant 8}$	$P_{\geqslant 1}$	$P_{\geqslant 2}$	Probability of interception by two profiles
0,05	0,334	0,004	0,008	0,012	0,016	0,020	0,025	0,031	0,551	0,666	0,663	
0,10	0,336	0,015	0,032	0,051	0,076	0,117	0,243	0,131		0,664	0,649	
0,15	0,339	0,035	0,077	0,144	0,318	0,087				0,661	0,626	
0,20	0,343	0,065	0,160	0,366	0,065					0,657	0,592	
0,25	0,349	0,107	0,343	0,201						0,651	0,544	
0,30	0,357	0,167	0,433	0,044						0,643	0,477	
0,35	0,365	0,272	0,363							0,635	0,363	
0,40	0,376	0,376	0,248							0,624	0,248	
0,45	0,388	0,448	0,164							0,612	0,164	
0,50	0,403	0,497	0,101							0,597	0,101	
0,60	0,440	0,538	0,022							0,560	0,022	
0,70	0,501	0,499								0,499		
0,80	0,564	0,436								0,436		
0,90	0,612	0,388								0,388		
1,00	0,651	0,349								0,349		
Continuous observation along the profile	0,333											

Table 17

b' = 1,0 d = 1,7

h in units of d	P_0	P_1	P_2	P_3	P_4	P_5	P_6	P_7	$P_{\geqslant 8}$	$P_{\geqslant 1}$	$P_{\geqslant 2}$	Probability of interception by two profiles
0,05	0,412	0,004	0,009	0,013	0,018	0,024	0,030	0,037	0,452	0,587	0,583	
0,10	0,415	0,017	0,036	0,060	0,095	0,191	0,185			0,585	0,568	
0,15	0,418	0,040	0,092	0,211	0,239					0,582	0,541	
0,20	0,423	0,075	0,221	0,281						0,577	0,502	
0,25	0,430	0,127	0,369	0,074						0,570	0,443	
0,30	0,438	0,217	0,344							0,562	0,344	
0,35	0,449	0,326	0,225							0,551	0,225	
0,40	0,461	0,399	0,140							0,539	0,140	
0,45	0,476	0,445	0,080							0,524	0,080	
0,50	0,494	0,469	0,037							0,506	0,037	
0,60	0,547	0,453								0,453		
0,70	0,612	0,388								0,388		
0,80	0,660	0,340								0,340		
0,90	0,698	0,302								0,302		
1,00	0,728	0,272								0,272		
Continuous observation along the profile	0,412											

Table 18

b' = 1,0 d = 2,0

h in units of d	P_0	P_1	P_2	P_3	P_4	P_5	P_6	P_7	$P_{\geqslant 8}$	$P_{\geqslant 1}$	$P_{\geqslant 2}$	Probability of interception by two profiles
0,05	0,501	0,005	0,010	0,016	0,022	0,029	0,038	0,049	0,330	0,499	0,494	
0,10	0,503	0,021	0,044	0,077	0,150	0,204	<0,0005			0,497	0,476	
0,15	0,508	0,049	0,120	0,275	0,049					0,492	0,444	
0,20	0,514	0,093	0,291	0,102						0,486	0,393	
0,25	0,522	0,171	0,307							0,478	0,307	
0,30	0,532	0,282	0,186							0,468	0,186	
0,35	0,545	0,350	0,106							0,455	0,106	
0,40	0,560	0,389	0,051							0,440	0,051	
0,45	0,580	0,404	0,016							0,420	0,016	
0,50	0,607	0,393								0,393		
0,60	0,673	0,327								0,327		
0,70	0,720	0,280								0,280		
0,80	0,755	0,245								0,245		
0,90	0,782	0,218								0,218		
1,00	0,804	0,196								0,196		
Continuous observation along the profile	0,500											

Table 19

$$b'=0,7 \qquad -30° \leqslant \Theta \leqslant 30° \qquad d=0,5$$

h in units of d	P_0	P_1	P_2	P_3	P_4	P_5	P_6	P_7	$P_{\geqslant 8}$	$P_{\geqslant 1}$	$P_{\geqslant 2}$	Probability of interception by two profiles
0,05									1,00	1,00	1,00	0,95
0,10									1,00	1,00	1,00	0,95
0,15									1,00	1,00	1,00	0,95
0,20							<0,005	0,04	0,96	1,00	1,00	0,95
0,25						0,01	0,06	0,09	0,84	1,00	1,00	0,94
0,30					0,01	0,07	0,11	0,30	0,50	1,00	1,00	0,94
0,35				<0,005	0,05	0,13	0,34	0,42	0,05	1,00	1,00	0,94
0,40				0,03	0,11	0,32	0,42	0,12	<0,005	1,00	1,00	0,93
0,45				0,06	0,20	0,52	0,21	<0,005		1,00	1,00	0,92
0,50			0,01	0,11	0,40	0,42	0,05			1,00	1,00	0,91
0,60			0,07	0,30	0,54	0,10	<0,005			1,00	1,00	0,90
0,70		<0,005	0,16	0,54	0,30	<0,005				1,00	1,00	0,87
0,80		0,02	0,38	0,44	0,16					1,00	0,98	0,85
0,90		0,04	0,55	0,33	0,08					1,00	0,96	0,82
1,00		0,08	0,66	0,23	0,03					1,00	0,92	0,78
Continuous observation along the profile	—											0,95

Table 20

$$b'=0,7 \qquad -90° \leqslant \Theta \leqslant 90° \qquad d=0,5$$

h in units of d	P_0	P_1	P_2	P_3	P_4	P_5	P_6	P_7	$P_{\geqslant 8}$	$P_{\geqslant 1}$	$P_{\geqslant 2}$	Probability of interception by two profiles
0,05									1,00	1,00	1,00	0,71
0,10									1,00	1,00	1,00	0,71
0,15									1,00	1,00	1,00	0,71
0,20							<0,005	0,02	0,98	1,00	1,00	0,71
0,25						<0,005	0,06	0,18	0,75	1,00	1,00	0,71
0,30					<0,005	0,08	0,23	0,20	0,49	1,00	1,00	0,70
0,35				<0,005	0,05	0,23	0,27	0,28	0,16	1,00	1,00	0,70
0,40				0,01	0,18	0,31	0,32	0,17	0,01	1,00	1,00	0,70
0,45				0,06	0,31	0,36	0,24	0,04	<0,005	1,00	1,00	0,69
0,50			<0,005	0,16	0,37	0,37	0,09	<0,005		1,00	1,00	0,69
0,60			0,05	0,41	0,37	0,17	<0,005			1,00	1,00	0,67
0,70		<0,005	0,18	0,53	0,26	0,03				1,00	1,00	0,66
0,80		0,01	0,37	0,48	0,14	<0,005				1,00	0,99	0,64
0,90		0,05	0,52	0,38	0,06					1,00	0,95	0,62
1,00		0,11	0,60	0,27	0,02					1,00	0,89	0,59
Continuous observation along the profile	—											0,71

Table 21

$$b'=0,7 \qquad -30° \leqslant \Theta \leqslant 30° \qquad d=0,6$$

h in units of d	P_0	P_1	P_2	P_3	P_4	P_5	P_6	P_7	$P_{\geqslant 8}$	$P_{\geqslant 1}$	$P_{>2}$	Probability of interception by two profiles
0,05									1,00	1,00	1,00	0,63
0,10									1,00	1,00	1,00	0,62
0,15								0,05	0,95	1,00	1,00	0,62
0,20						0,04	0,34	0,07	0,55	1,00	1,00	0,62
0,25					0,10	0,31	0,13	0,30	0,16	1,00	1,00	0,62
0,30				0,03	0,37	0,19	0,30	0,11	<0,005	1,00	1,00	0,61
0,35				0,25	0,27	0,37	0,12	<0,005		1,00	1,00	0,61
0,40			0,02	0,42	0,30	0,25	0,01			1,00	1,00	0,60
0,45			0,15	0,40	0,34	0,10	<0,005			1,00	1,00	0,59
0,50			0,27	0,43	0,28	0,02				1,00	1,00	0,58
0,60		0,02	0,53	0,36	0,10					1,00	0,98	0,56
0,70		0,14	0,58	0,25	0,04					1,00	0,86	0,53
0,80		0,25	0,60	0,14	0,01					1,00	0,75	0,50
0,90		0,36	0,58	0,06	<0,005					1,00	0,64	0,46
1,00	<0,005	0,48	0,51	0,01						1,00	0,52	0,41
Continuous observation along the profile	—											0,63

Table 22

$$b'=07 \qquad -90° \leqslant \Theta \leqslant 90° \qquad d=0,6$$

h in units of d	P_0	P_1	P_2	P_3	P_4	P_5	P_6	P_7	$P_{\geqslant 8}$	$P_{\geqslant 1}$	$P_{>2}$	Probability of interception by two profiles
0,05									1,00	1,00	1,00	0,43
0,10									1,00	1,00	1,00	0,42
0,15								0,02	0,98	1,00	1,00	0,42
0,20						0,01	0,22	0,26	0,51	1,00	1,00	0,42
0,25					0,04	0,30	0,30	0,24	0,12	1,00	1,00	0,42
0,30				0,01	0,29	0,37	0,26	0,07	<0,005	1,00	1,00	0,42
0,35				0,14	0,45	0,30	0,10	<0,005		1,00	1,00	0,41
0,40			0,01	0,37	0,44	0,17	0,01			1,00	1,00	0,41
0,45			0,08	0,53	0,32	0,08	<0,005			1,00	1,00	0,40
0,50			0,19	0,58	0,21	0,02				1,00	1,00	0,39
0,60		0,01	0,50	0,44	0,05					1,00	0,99	0,38
0,70		0,09	0,64	0,25	0,01					1,00	0,91	0,36
0,80		0,22	0,65	0,13	<0,005					1,00	0,78	0,34
0,90	<0,005	0,36	0,58	0,06	<0,005					1,00	0,64	0,31
1,00	<0,005	0,49	0,49	0,02						1,00	0,51	0,28
Continuous observation along the profile	—											0,43

Table 23

b′=0,7   −30° ≤ Θ ≤ 30°   d=0,7

h in units of d	P_0	P_1	P_2	P_3	P_4	P_5	P_6	P_7	$P_{\geqslant 8}$	$P_{\geqslant 1}$	$P_{\geqslant 2}$	Probability of interception by two profiles
0,05									1,00	1,00	1,00	0,39
0,10									1,00	1,00	1,00	0,39
0,15							0,24	0,39	0,36	1,00	1,00	0,39
0,20					0,06	0,55	0,16	0,19	0,04	1,00	1,00	0,39
0,25				0,05	0,58	0,21	0,15	0,01		1,00	1,00	0,38
0,30				0,45	0,37	0,17	0,01			1,00	1,00	0,38
0,35			0,11	0,62	0,21	0,05	<0,005			1,00	1,00	0,37
0,40			0,34	0,51	0,14	<0,005				1,00	1,00	0,36
0,45		<0,005	0,56	0,38	0,06					1,00	1,00	0,35
0,50		0,03	0,72	0,22	0,03					1,00	0,97	0,34
0,60		0,25	0,64	0,11	<0,005					1,00	0,75	0,31
0,70		0,43	0,53	0,04	<0,005					1,00	0,57	0,28
0,80	<0,005	0,60	0,40	<0,005						1,00	0,40	0,24
0,90	0,02	0,72	0,26							0,98	0,26	0,20
1,00	0,07	0,75	0,19							0,93	0,19	0,18
Continuous observation along the profile	—											0,40

Table 24

b′=0,7   −90° ≤ Θ ≤ 90°   d=0,7

h in units of d	P_0	P_1	P_2	P_3	P_4	P_5	P_6	P_7	$P_{\geqslant 8}$	$P_{\geqslant 1}$	$P_{\geqslant 2}$	Probability of interception by two profiles
0,05	<0,005	<0,005	<0,005	<0,005	<0,005	<0,005	<0,005	<0,005	1,00	1,00	1,00	0,22
0,10	<0,005	<0,005	<0,005	<0,005	<0,005	0,01	0,01	0,01	0,97	1,00	1,00	0,22
0,15	<0,005	<0,005	<0,005	0,01	0,01	0,03	0,16	0,32	0,47	1,00	1,00	0,22
0,20	<0,005	<0,005	0,01	0,02	0,07	0,39	0,30	0,20	0,02	1,00	1,00	0,22
0,25	<0,005	<0,005	0,01	0,07	0,44	0,35	0,12	<0,005		1,00	1,00	0,21
0,30	<0,005	0,01	0,03	0,34	0,48	0,14	0,01			1,00	0,99	0,21
0,35	<0,005	0,01	0,11	0,56	0,30	0,02	<0,005			1,00	0,99	0,20
0,40	<0,005	0,02	0,28	0,58	0,12	<0,005				1,00	0,98	0,20
0,45	<0,005	0,03	0,48	0,46	0,03					1,00	0,97	0,19
0,50	<0,005	0,06	0,64	0,29	0,01					1,00	0,94	0,19
0,60	<0,005	0,22	0,68	0,10	<0,005					1,00	0,77	0,17
0,70	0,01	0,41	0,56	0,02	<0,005					0,99	0,59	0,15
0,80	0,01	0,58	0,41	<0,005						0,99	0,41	0,12
0,90	0,02	0,71	0,27							0,98	0,27	0,10
1,00	0,05	0,78	0,17							0,95	0,17	0,09
Continuous observation along the profile	—											0,22

24

Table 25

$$b'=0{,}7 \qquad -30° \leqslant \Theta \leqslant 30° \qquad d=0{,}8$$

h in units of d	P_0	P_1	P_2	P_3	P_4	P_5	P_6	P_7	$P_{\geqslant 8}$	$P_{\geqslant 1}$	$P_{\geqslant 2}$	Probability of interception by two profiles
0,05									1,00	1,00	1,00	0,22
0,10							<0,005	0,12	0,88	1,00	1,00	0,22
0,15					0,02	0,35	0,51	0,10	0,01	1,00	1,00	0,22
0,20				0,06	0,61	0,30	0,03			1,00	1,00	0,21
0,25			0,02	0,56	0,39	0,03	<0,005			1,00	1,00	0,21
0,30			0,20	0,73	0,07	<0,005				1,00	1,00	0,20
0,35		<0,005	0,55	0,43	0,02					1,00	1,00	0,19
0,40		0,05	0,76	0,18	0,01					1,00	0,95	0,18
0,45		0,15	0,79	0,06	<0,005					1,00	0,85	0,17
0,50		0,31	0,66	0,03	<0,005					1,00	0,69	0,15
0,60	<0,005	0,57	0,43	<0,005						1,00	0,43	0,12
0,70	0,02	0,74	0,24							0,98	0,24	0,10
0,80	0,06	0,81	0,13							0,94	0,13	0,09
0,90	0,12	0,80	0,08							0,88	0,08	0,08
1,00	0,21	0,72	0,07							0,79	0,07	0,07
Continuous observation along the profile	—											0,22

Table 26

$$b'=0{,}7 \qquad -90° \leqslant \Theta \leqslant 90° \qquad d=0{,}8$$

h in units of d	P_0	P_1	P_2	P_3	P_4	P_5	P_6	P_7	$P_{\geqslant 8}$	$P_{\geqslant 1}$	$P_{\geqslant 2}$	Probability of interception by two profiles
0,05	0,03	<0,005	<0,005	<0,005	<0,005	<0,005	<0,005	0,01	0,95	0,97	0,97	0,10
0,10	0,03	<0,005	0,01	0,01	0,01	0,02	0,04	0,10	0,79	0,97	0,97	0,10
0,15	0,03	0,01	0,01	0,03	0,06	0,22	0,35	0,20	0,10	0,97	0,96	0,10
0,20	0,03	0,01	0,03	0,10	0,37	0,34	0,13	<0,005		0,97	0,96	0,09
0,25	0,03	0,02	0,07	0,36	0,41	0,11	<0,005			0,97	0,95	0,09
0,30	0,03	0,03	0,19	0,54	0,20	<0,005				0,97	0,94	0,09
0,35	0,03	0,05	0,40	0,47	0,05					0,97	0,92	0,08
0,40	0,04	0,09	0,57	0,30	< 0,005					0,96	0,88	0,08
0,45	0,04	0,16	0,66	0,15	<0,005					0,96	0,80	0,07
0,50	0,04	0,26	0,64	0,06	<0,005					0,96	0,70	0,07
0,60	0,05	0,47	0,48	<0,005						0,95	0,48	0,05
0,70	0,06	0,65	0,29							0,94	0,29	0,04
0,80	0,09	0,74	0,17							0,91	0,17	0,03
0,90	0,13	0,78	0,09							0,87	0,09	0,03
1,00	0,19	0,76	0,05							0,81	0,05	0,03
Continuous observation along the profile	0,03											0,10

Table 27

$$b'=0,7 \quad -30° \leqslant \Theta \leqslant 30° \quad d=0,9$$

h in units of d	P_0	P_1	P_2	P_3	P_4	P_5	P_6	P_7	$P_{\geqslant 8}$	$P_{\geqslant 1}$	$P_{\geqslant 2}$	Probability of interception by two profiles
0,05							<0,005	<0,005	1,00	1,00	1,00	0,08
0,10				<0,005	0,03	0,12	0,20	0,31	0,33	1,00	1,00	0,08
0,15			0,01	0,12	0,30	0,50	0,08			1,00	1,00	0,08
0,20		<0,005	0,10	0,41	0,49	<0,005				1,00	1,00	0,07
0,25	<0,005	0,02	0,29	0,64	0,05					1,00	0,98	0,07
0,30	<0,005	0,07	0,59	0,33	<0,005					1,00	0,93	0,06
0,35	<0,005	0,16	0,74	0,10						1,00	0,84	0,05
0,40	<0,005	0,30	0,69	<0,005						1,00	0,70	0,04
0,45	0,01	0,48	0,51							0,99	0,51	0,03
0,50	0,02	0,61	0,38							0,98	0,38	0,03
0,60	0,05	0,77	0,18							0,95	0,18	0,02
0,70	0,10	0,83	0,07							0,90	0,07	0,02
0,80	0,17	0,81	0,02							0,83	0,02	0,02
0,90	0,26	0,72	0,02							0,74	0,02	0,02
1,00	0,33	0,65	0,01							0,67	0,01	0,01
Continuous observation along the profile	—											0.09

Table 28

$$b'=0,7 \quad -90° \leqslant \Theta \leqslant 90° \quad d=0,9$$

h in units of d	P_0	P_1	P_2	P_3	P_4	P_5	P_6	P_7	$P_{\geqslant 8}$	$P_{\geqslant 1}$	$P_{\geqslant 2}$	Probability of interception by two profiles
0,05	0,08	<0,005	<0,005	<0,005	0,01	0,01	0,01	0,01	0,88	0,92	0,92	0,03
0,10	0,08	0,01	0,01	0,02	0,04	0,08	0,12	0,17	0,47	0,92	0,92	0,03
0,15	0,08	0,01	0,03	0,09	0,18	0,30	0,20	0,10	0,01	0,92	0,91	0,03
0,20	0,08	0,03	0,09	0,24	0,37	0,17	0,02			0,92	0,89	0,03
0,25	0,08	0,05	0,20	0,43	0,22	0,02				0,92	0,87	0,02
0,30	0,09	0,09	0,37	0,40	0,06					0,91	0,83	0,02
0,35	0,09	0,14	0,51	0,25	<0,005					0,91	0,77	0,02
0,40	0,09	0,22	0,57	0,11						0,91	0,68	0,01
0,45	0,10	0,33	0,53	0,04						0,90	0,57	0,01
0,50	0,11	0,43	0,45	0,01						0,89	0,46	0,01
0,60	0,13	0,60	0,27							0,87	0,27	0,01
0,70	0,17	0,70	0,14							0,83	0,14	0,01
0,80	0,21	0,73	0,06							0,79	0,06	0,01
0,90	0,27	0,70	0,02							0,73	0,02	0,01
1,00	0,33	0,66	0,01							0,67	0,01	<0,005
Continuous observation along the profile	0,08											0,03

Table 29

$$b' = 0,7 \qquad -30° \leqslant \Theta \leqslant 30° \qquad d = 1,0$$

h in units of d	P_0	P_1	P_2	P_3	P_4	P_5	P_6	P_7	$P_{\geqslant 8}$	$P_{\geqslant 1}$	$P_{\geqslant 2}$	Probability of interception by two profiles
0,05	0,023	0,005	0,010	0,015	0,020	0,026	0,032	0,038	0,832	0,977	0,972	
0,10	0,026	0,019	0,040	0,064	0,093	0,136	0,229	0,364	0,028	0,974	0,955	
0,15	0,030	0,044	0,096	0,171	0,377	0,281				0,970	0,926	
0,20	0,035	0,081	0,193	0,479	0,211					0,965	0,883	
0,25	0,043	0,133	0,407	0,417						0,957	0,824	
0,30	0,052	0,204	0,603	0,140						0,948	0,744	
0,35	0,063	0,311	0,617	0,008						0,937	0,625	
0,40	0,076	0,473	0,450							0,924	0,450	
0,45	0,092	0,595	0,313							0,909	0,313	
0,50	0,109	0,682	0,209							0,891	0,209	
0,60	0,154	0,776	0,070							0,846	0,070	
0,70	0,219	0,777	0,004							0,781	0,004	
0,80	0,313	0,687								0,687		
0,90	0,389	0,611								0,611		
1,00	0,450	0,550								0,550		
Continuous observation along the profile	0,023											

Table 30

$$b' = 0,7 \qquad -90° \leqslant \Theta \leqslant 90° \qquad d = 1,0$$

h in units of d	P_0	P_1	P_2	P_3	P_4	P_5	P_6	P_7	$P_{\geqslant 8}$	$P_{\geqslant 1}$	$P_{\geqslant 2}$	Probability of interception by two profiles
0,05	0,144	0,003	0,007	0,010	0,014	0,018	0,022	0,026	0,756	0,856	0,853	
0,10	0,146	0,014	0,028	0,044	0,063	0,089	0,138	0,221	0,257	0,854	0,841	
0,15	0,149	0,031	0,066	0,114	0,221	0,265	0,130	0,025		0,852	0,820	
0,20	0,152	0,057	0,130	0,293	0,287	0,081				0,847	0,791	
0,25	0,158	0,092	0,253	0,391	0,108					0,842	0,751	
0,30	0,164	0,139	0,410	0,275	0,012					0,836	0,697	
0,35	0,172	0,206	0,503	0,120						0,828	0,622	
0,40	0,181	0,305	0,474	0,041						0,819	0,515	
0,45	0,191	0,404	0,396	0,008						0,809	0,405	
0,50	0,203	0,494	0,303							0,797	0,303	
0,60	0,234	0,617	0,150							0,767	0,150	
0,70	0,275	0,665	0,060							0,725	0,060	
0,80	0,333	0,647	0,020							0,667	0,020	
0,90	0,393	0,603	0,004							0,607	0,004	
1,00	0,450	0,550								0,550		
Continuous observation along the profile	0,143											

Table 31

b′=0,7   —30° ⩽ Θ ⩽ 30°   d=1,1

h in units of d	P_0	P_1	P_2	P_3	P_4	P_5	P_6	P_7	$P_{\geqslant 8}$	$P_{\geqslant 1}$	$P_{\geqslant 2}$	Probability of interception by two profiles
0,05	0,112	0,005	0,011	0,016	0,022	0,028	0,036	0,044	0,726	0,888	0,883	
0,10	0,115	0,021	0,044	0,072	0,108	0,171	0,347	0,121		0,885	0,864	
0,15	0,119	0,049	0,109	0,208	0,434	0,080				0,881	0,831	
0,20	0,126	0,091	0,230	0,493	0,060					0,875	0,784	
0,25	0,134	0,150	0,480	0,235						0,866	0,716	
0,30	0,144	0,238	0,578	0,040						0,856	0,618	
0,35	0,156	0,389	0,454							0,844	0,454	
0,40	0,171	0,522	0,307							0,829	0,307	
0,45	0,188	0,613	0,198							0,812	0,198	
0,50	0,209	0,673	0,118							0,791	0,118	
0,60	0,263	0,717	0,020							0,737	0,020	
0,70	0,351	0,649								0,649		
0,80	0,432	0,568								0,568		
0,90	0,495	0,505								0,505		
1,00	0,546	0,454								0,454		
Continuous observation along the profile	0,111											

Table 32

b′=0,7   —90° ⩽ Θ ⩽ 90°   d=1,1

h in units of d	P_0	P_1	P_2	P_3	P_4	P_5	P_6	P_7	$P_{\geqslant 8}$	$P_{\geqslant 1}$	$P_{\geqslant 2}$	Probability of interception by two profiles
0,05	0,222	0,004	0,007	0,011	0,015	0,020	0,025	0,030	0,666	0,778	0,774	
0,10	0,224	0,015	0,031	0,049	0,073	0,109	0,196	0,164	0,139	0,777	0,762	
0,15	0,227	0,034	0,075	0,136	0,271	0,191	0,066	<0,0005		0,773	0,739	
0,20	0,231	0,063	0,152	0,333	0,199	0,022				0,769	0,706	
0,25	0,237	0,103	0,305	0,314	0,040					0,763	0,660	
0,30	0,244	0,160	0,434	0,162	<0,0005					0,756	0,596	
0,35	0,253	0,249	0,446	0,052						0,747	0,498	
0,40	0,263	0,350	0,376	0,011						0,737	0,388	
0,45	0,275	0,441	0,284	<0,0005						0,725	0,284	
0,50	0,289	0,514	0,197							0,711	0,197	
0,60	0,324	0,595	0,081							0,676	0,081	
0,70	0,377	0,597	0,026							0,623	0,026	
0,80	0,438	0,557	0,006							0,562	0,006	
0,90	0,495	0,505	<0,0005							0,505	<0,0005	
1,00	0,546	0,454								0,454		
Continuous observation along the profile	0,221											

28

Table 33

$$b' = 0,7 \qquad -30° \leqslant \Theta \leqslant 30° \qquad d = 1,2$$

h in units of d	P_0	P_1	P_2	P_3	P_4	P_5	P_6	P_7	$P_{\geqslant 8}$	$P_{\geqslant 1}$	$P_{\geqslant 2}$	Probability of interception by two profiles
0,05	0,186	0,006	0,012	0,018	0,024	0,032	0,040	0,050	0,633	0,814	0,808	
0,10	0,189	0,023	0,049	0,081	0,127	0,243	0,283	0,005		0,811	0,787	
0,15	0,194	0,054	0,123	0,273	0,353	0,003				0,806	0,752	
0,20	0,201	0,101	0,288	0,407	0,002					0,799	0,698	
0,25	0,210	0,170	0,503	0,117						0,790	0,620	
0,30	0,221	0,287	0,491	0,002						0,779	0,492	
0,35	0,235	0,439	0,326							0,765	0,326	
0,40	0,251	0,543	0,206							0,749	0,206	
0,45	0,271	0,609	0,120							0,729	0,120	
0,50	0,295	0,646	0,059							0,705	0,059	
0,60	0,364	0,635	0,001							0,636	0,001	
0,70	0,455	0,545								0,545		
0,80	0,523	0,477								0,477		
0,90	0,576	0,424								0,424		
1,00	0,618	0,382								0,382		
Continuous observation along the profile	0,185											

Table 34

$$b' = 0,7 \qquad -90° \leqslant \Theta \leqslant 90° \qquad d = 1,2$$

h in units of d	P_0	P_1	P_2	P_3	P_4	P_5	P_6	P_7	$P_{\geqslant 8}$	$P_{\geqslant 1}$	$P_{\geqslant 2}$	Probability of interception by two profiles
0,05	0,287	0,004	0,008	0,012	0,017	0,022	0,027	0,034	0,589	0,713	0,709	
0,10	0,289	0,016	0,034	0,055	0,084	0,143	0,203	0,111	0,065	0,711	0,695	
0,15	0,292	0,038	0,084	0,168	0,272	0,129	0,017			0,708	0,670	
0,20	0,297	0,070	0,183	0,330	0,119	0,002				0,703	0,633	
0,25	0,303	0,116	0,342	0,229	0,010					0,697	0,581	
0,30	0,311	0,187	0,420	0,082						0,689	0,502	
0,35	0,321	0,288	0,372	0,020						0,679	0,392	
0,40	0,332	0,383	0,284	0,001						0,668	0,285	
0,45	0,345	0,461	0,194							0,655	0,194	
0,50	0,361	0,514	0,125							0,639	0,125	
0,60	0,405	0,555	0,041							0,595	0,041	
0,70	0,464	0,526	0,010							0,536	0,010	
0,80	0,523	0,476	<0,0005							0,477	<0,0005	
0,90	0,576	0,424								0,424		
1,00	0,618	0,382								0,382		
Continuous observation along the profile	0,286											

Table 35

$b'=0{,}7$ $-30° \leqslant \Theta \leqslant 30°$ $d=1{,}3$

h in units of d	P_0	P_1	P_2	P_3	P_4	P_5	P_6	P_7	$P_{\geqslant 8}$	$P_{\geqslant 1}$	$P_{\geqslant 2}$	Probability of interception by two profiles
0,05	0,249	0,006	0,013	0,019	0,027	0,035	0,045	0,057	0,550	0,751	0,745	
0,10	0,252	0,025	0,054	0,090	0,151	0,318	0,110			0,748	0,722	
0,15	0,257	0,059	0,139	0,345	0,199					0,743	0,683	
0,20	0,265	0,112	0,355	0,268						0,735	0,623	
0,25	0,275	0,193	0,488	0,044						0,725	0,532	
0,30	0,287	0,341	0,372							0,713	0,372	
0,35	0,302	0,466	0,232							0,698	0,232	
0,40	0,321	0,545	0,134							0,679	0,134	
0,45	0,343	0,590	0,066							0,657	0,066	
0,50	0,371	0,607	0,022							0,629	0,022	
0,60	0,458	0,542								0,542		
0,70	0,535	0,465								0,465		
0,80	0,593	0,407								0,407		
0,90	0,639	0,361								0,361		
1,00	0,675	0,325								0,325		
Continuous observation along the profile	0,248											

Table 36

$b'=0{,}7$ $-90° \leqslant \Theta \leqslant 90°$ $d=1{,}3$

h in units of d	P_0	P_1	P_2	P_3	P_4	P_5	P_6	P_7	$P_{\geqslant 8}$	$P_{\geqslant 1}$	$P_{\geqslant 2}$	Probability of interception by two profiles
0,05	0,342	0,004	0,009	0,014	0,019	0,024	0,030	0,038	0,521	0,658	0,654	
0,10	0,344	0,018	0,037	0,061	0,098	0,184	0,157	0,084	0,018	0,656	0,638	
0,15	0,348	0,041	0,094	0,208	0,231	0,077	0,001			0,653	0,611	
0,20	0,353	0,077	0,220	0,290	0,059					0,647	0,570	
0,25	0,360	0,130	0,360	0,150	0,001					0,640	0,510	
0,30	0,368	0,219	0,373	0,040						0,632	0,413	
0,35	0,379	0,318	0,298	0,005						0,621	0,303	
0,40	0,391	0,404	0,205							0,609	0,205	
0,45	0,406	0,464	0,129							0,594	0,129	
0,50	0,425	0,500	0,075							0,575	0,075	
0,60	0,478	0,503	0,020							0,522	0,020	
0,70	0,538	0,460	0,003							0,462	0,003	
0,80	0,593	0,407								0,407		
0,90	0,639	0,361								0,361		
1,00	0,675	0,325								0,325		
Continuous observation along the profile	0,341											

Table 37

$$b'=0,7 \qquad -30° \leqslant \Theta \leqslant 30° \qquad d=1,4$$

h in units of d	P_0	P_1	P_2	P_3	P_4	P_5	P_6	P_7	$P_{\geqslant 8}$	$P_{\geqslant 1}$	$P_{\geqslant 2}$	Probability of interception by two profiles
0,05	0,303	0,007	0,014	0,021	0,029	0,038	0,050	0,065	0,474	0,697	0,690	
0,10	0,306	0,028	0,059	0,101	0,189	0,303	0,015			0,694	0,666	
0,15	0,312	0,065	0,158	0,373	0,093					0,688	0,623	
0,20	0,320	0,123	0,391	0,166						0,680	0,557	
0,25	0,331	0,222	0,441	0,006						0,669	0,447	
0,30	0,344	0,376	0,279							0,656	0,279	
0,35	0,361	0,477	0,162							0,639	0,162	
0,40	0,382	0,535	0,083							0,618	0,083	
0,45	0,408	0,562	0,031							0,593	0,031	
0,50	0,442	0,555	0,003							0,558	0,003	
0,60	0,533	0,467								0,467		
0,70	0,599	0,401								0,401		
0,80	0,649	0,351								0,351		
0,90	0,688	0,312								0,312		
1,00	0,719	0,280								0,280		
Continuous observation along the profile	0,302											

Table 38

$$b'=0,7 \qquad -90° \leqslant \Theta \leqslant 90° \qquad d=1,4$$

h in units of d	P_0	P_1	P_2	P_3	P_4	P_5	P_6	P_7	$P_{\geqslant 8}$	$P_{\geqslant 1}$	$P_{\geqslant 2}$	Probability of interception by two profiles
0,05	0,389	0,005	0,010	0,015	0,020	0,026	0,034	0,043	0,459	0,611	0,607	
0,10	0,391	0,019	0,041	0,068	0,117	0,200	0,113	0,049	0,001	0,609	0,589	
0,15	0,395	0,045	0,105	0,238	0,183	0,034				0,605	0,560	
0,20	0,401	0,085	0,251	0,238	0,025					0,599	0,515	
0,25	0,408	0,147	0,359	0,086						0,592	0,445	
0,30	0,418	0,247	0,319	0,017						0,582	0,336	
0,35	0,429	0,340	0,230	$<0,0005$						0,571	0,230	
0,40	0,443	0,412	0,144							0,557	0,144	
0,45	0,460	0,456	0,084							0,540	0,084	
0,50	0,482	0,475	0,043							0,518	0,043	
0,60	0,541	0,451	0,008							0,459	0,008	
0,70	0,599	0,401	$<0,0005$							0,401	$<0,0005$	
0,80	0,649	0,351								0,351		
0,90	0,688	0,312								0,312		
1,00	0,720	0,280								0,280		
Continuous observation along the profile	0,388											

Table 39

b′=0,7   —30° ≤ Θ ≤ 30°   d=1,5

h in units of d	P_0	P_1	P_2	P_3	P_4	P_5	P_6	P_7	$P_{\geqslant 8}$	$P_{\geqslant 1}$	$P_{\geqslant 2}$	Probability of interception by two profiles
0,05	0,350	0,007	0,015	0,023	0,032	0,042	0,056	0,075	0,401	0,651	0,643	
0,10	0,353	0,030	0,064	0,114	0,251	0,187				0,647	0,617	
0,15	0,359	0,070	0,181	0,362	0,028					0,641	0,571	
0,20	0,368	0,136	0,402	0,094						0,632	0,496	
0,25	0,380	0,264	0,357							0,621	0,357	
0,30	0,394	0,397	0,209							0,606	0,209	
0,35	0,413	0,476	0,111							0,587	0,111	
0,40	0,436	0,517	0,047							0,564	0,047	
0,45	0,466	0,525	0,009							0,534	0,009	
0,50	0,511	0,489								0,489		
0,60	0,593	0,407								0,407		
0,70	0,651	0,349								0,349		
0,80	0,695	0,305								0,305		
0,90	0,729	0,271								0,271		
1,00	0,756	0,244								0,244		
Continuous observation along the profile	0,348											

Table 40

b′=0,7   —90° ≤ Θ ≤ 90°   d=1,5

h in units of d	P_0	P_1	P_2	P_3	P_4	P_5	P_6	P_7	$P_{\geqslant 8}$	$P_{\geqslant 1}$	$P_{\geqslant 2}$	Probability of interception by two profiles
0,05	0,430	0,005	0,010	0,016	0,022	0,029	0,037	0,049	0,402	0,570	0,565	
0,10	0,432	0,021	0,044	0,076	0,147	0,176	0,087	0,016		0,568	0,547	
0,15	0,437	0,049	0,119	0,253	0,133	0,011				0,564	0,515	
0,20	0,443	0,093	0,273	0,183	0,008					0,557	0,465	
0,25	0,451	0,169	0,332	0,048						0,549	0,380	
0,30	0,461	0,269	0,264	0,005						0,539	0,270	
0,35	0,473	0,355	0,172							0,527	0,172	
0,40	0,489	0,411	0,100							0,511	0,100	
0,45	0,508	0,440	0,051							0,491	0,051	
0,50	0,535	0,441	0,024							0,465	0,024	
0,60	0,595	0,402	0,003							0,405	0,003	
0,70	0,651	0,349								0,349		
0,80	0,695	0,305								0,305		
0,90	0,729	0,271								0,271		
1,00	0,756	0,244								0,244		
Continuous observation along the profile	0,429											

Table 41

b′=0,7 —30° ≤ Θ ≤ 30° d=1,7

h in units of d	P_0	P_1	P_2	P_3	P_4	P_5	P_6	P_7	$P_{\geqslant 8}$	$P_{\geqslant 1}$	$P_{\geqslant 2}$	Probability of interception by two profiles
0,05	0,426	0,008	0,017	0,026	0,037	0,051	0,071	0,107	0,257	0,574	0,566	
0,10	0,430	0,034	0,076	0,150	0,283	0,027				0,570	0,536	
0,15	0,437	0,082	0,256	0,225						0,563	0,481	
0,20	0,447	0,167	0,372	0,013						0,553	0,385	
0,25	0,461	0,317	0,222							0,539	0,222	
0,30	0,478	0,409	0,112							0,522	0,112	
0,35	0,501	0,455	0,044							0,499	0,044	
0,40	0,531	0,462	0,007							0,469	0,007	
0,45	0,577	0,423								0,423		
0,50	0,620	0,380								0,380		
0,60	0,683	0,317								0,317		
0,70	0,728	0,272								0,272		
0,80	0,762	0,238								0,238		
0,90	0,789	0,211								0,211		
1,00	0,810	0,190								0,190		
Continuous observation along the profile	0,425											

Table 42

b′=0,7 —90° ≤ Θ ≤ 90° d=1,7

h in units of d	P_0	P_1	P_2	P_3	P_4	P_5	P_6	P_7	$P_{\geqslant 8}$	$P_{\geqslant 1}$	$P_{\geqslant 2}$	Probability of interception by two profiles
0,05	0,497	0,006	0,012	0,018	0,025	0,034	0,046	0,066	0,296	0,503	0,497	
0,10	0,500	0,024	0,052	0,096	0,185	0,113	0,030			0,500	0,477	
0,15	0,505	0,056	0,158	0,227	0,054					0,495	0,439	
0,20	0,512	0,112	0,290	0,086						0,488	0,376	
0,25	0,521	0,209	0,258	0,012						0,479	0,270	
0,30	0,533	0,300	0,167							0,467	0,167	
0,35	0,548	0,360	0,092							0,452	0,092	
0,40	0,568	0,389	0,043							0,432	0,043	
0,45	0,595	0,387	0,018							0,405	0,018	
0,50	0,626	0,368	0,006							0,374	0,006	
0,60	0,683	0,317								0,317		
0,70	0,728	0,272								0,272		
0,80	0,762	0,238								0,238		
0,90	0,789	0,211								0,211		
1,00	0,810	0,190								0,190		
Continuous observation along the profile	0,496											

Table 43

$b'=0,7$ $\quad$ $-30° \leqslant \Theta \leqslant 30°$ $\quad$ $d=2,0$

h in units of d	P_0	P_1	P_2	P_3	P_4	P_5	P_6	P_7	$P_{\geqslant 8}$	$P_{\geqslant 1}$	$P_{\geqslant 2}$	Probability of interception by two profiles
0,05	0,513	0,010	0,020	0,032	0,047	0,068	0,014	0,182	0,014	0,487	0,477	
0,10	0,518	0,041	0,096	0,240	0,105					0,482	0,442	
0,15	0,526	0,102	0,302	0,070						0,474	0,372	
0,20	0,538	0,237	0,225							0,462	0,225	
0,25	0,555	0,341	0,104							0,445	0,104	
0,30	0,577	0,388	0,035							0,423	0,035	
0,35	0,609	0,389	0,002							0,391	0,002	
0,40	0,656	0,344								0,344		
0,45	0,695	0,305								0,305		
0,50	0,725	0,275								0,275		
0,60	0,771	0,229								0,229		
0,70	0,804	0,196								0,196		
0,80	0,828	0,172								0,172		
0,90	0,847	0,153								0,153		
1,00	0,863	0,137								0,137		
Continuous observation along the profile	0,511											

Table 44

$b'=0,7$ $\quad$ $-90° \leqslant \Theta \leqslant 90°$ $\quad$ $d=2,0$

h in units of d	P_0	P_1	P_2	P_3	P_4	P_5	P_6	P_7	$P_{\geqslant 8}$	$P_{\geqslant 1}$	$P_{\geqslant 2}$	Probability of interception by two profiles
0,05	0,573	0,007	0,014	0,022	0,032	0,045	0,069	0,111	0,129	0,427	0,421	
0,10	0,576	0,028	0,065	0,146	0,144	0,041	<0,0005			0,424	0,396	
0,15	0,582	0,069	0,205	0,137	0,006					0,418	0,349	
0,20	0,590	0,152	0,237	0,020						0,410	0,257	
0,25	0,602	0,247	0,151							0,398	0,151	
0,30	0,617	0,308	0,075							0,383	0,075	
0,35	0,637	0,333	0,030							0,363	0,030	
0,40	0,667	0,323	0,010							0,333	0,010	
0,45	0,697	0,301	0,002							0,303	0,002	
0,50	0,725	0,275								0,275		
0,60	0,771	0,229								0,229		
0,70	0,804	0,196								0,196		
0,80	0,828	0,172								0,172		
0,90	0,847	0,153								0,153		
1,00	0,863	0,137								0,137		
Continuous observation along the profile	0,572											

Table 45

b′=0,5   —30° ⩽ Θ ⩽ 30°   d=0,5

h in units of d	P_0	P_1	P_2	P_3	P_4	P_5	P_6	P_7	$P_{\geqslant 8}$	$P_{\geqslant 1}$	$P_{\geqslant 2}$	Probability of interception by two profiles
0,05									1,00	1,00	1,00	0,93
0,10									1,00	1,00	1,00	0,93
0,15							<0,005	0,07	0,93	1,00	1,00	0,93
0,20					<0,005	0,06	0,09	0,17	0,69	1,00	1,00	0,92
0,25				<0,005	0,07	0,13	0,32	0,41	0,07	1,00	1,00	0,91
0,30				0,04	0,16	0,36	0,40	0,04		1,00	1,00	0,91
0,35			<0,005	0,12	0,35	0,45	0,08			1,00	1,00	0,90
0,40			0,04	0,23	0,49	0,23	0,01			1,00	1,00	0,88
0,45			0,09	0,40	0,46	0,06				1,00	1,00	0,87
0,50		<0,005	0,15	0,56	0,29	<0,005				1,00	1,00	0,85
0,60		0,05	0,35	0,52	0,07					1,00	0,95	0,82
0,70		0,12	0,53	0,34	0,02					1,00	0,88	0,77
0,80		0,20	0,64	0,16	<0,005					1,00	0,80	0,72
0,90	<0,005	0,30	0,64	0,05						1,00	0,70	0,65
1,00	0,01	0,41	0,57	0,01						0,99	0,58	0,57
Continuous observation along the profile	—											0,93

Table 46

b′=0,5   —90° ⩽ Θ ⩽ 90°   d=0,5

h in units of d	P_0	P_1	P_2	P_3	P_4	P_5	P_6	P_7	$P_{\geqslant 8}$	$P_{\geqslant 1}$	$P_{\geqslant 2}$	Probability of interception by two profiles
0,05	<0,005	<0,005	<0,005	<0,005	<0,005	<0,005	<0,005	<0,005	1,00	1,00	1,00	0,54
0,10	<0,005	<0,005	<0,005	<0,005	<0,005	<0,005	<0,005	<0,005	0,99	1,00	1,00	0,54
0,15	<0,005	<0,005	<0,005	<0,005	<0,005	<0,005	0,01	0,04	0,94	1,00	1,00	0,54
0,20	<0,005	<0,005	<0,005	<0,005	0,01	0,04	0,13	0,20	0,62	1,00	1,00	0,53
0,25	<0,005	<0,005	<0,005	0,01	0,05	0,19	0,29	0,32	0,14	1,00	1,00	0,53
0,30	<0,005	<0,005	<0,005	0,03	0,19	0,34	0,34	0,09	<0,005	1,00	1,00	0,53
0,35	<0,005	<0,005	0,01	0,12	0,36	0,39	0,12	<0,005		1,00	1,00	0,52
0,40	<0,005	<0,005	0,03	0,26	0,45	0,24	0,01			1,00	1,00	0,51
0,45	<0,005	<0,005	0,08	0,41	0,42	0,08	<0,005			1,00	1,00	0,51
0,50	<0,005	0,01	0,16	0,53	0,29	0,01				1,00	0,99	0,50
0,60	<0,005	0,03	0,40	0,49	0,08					1,00	0,97	0,48
0,70	<0,005	0,10	0,58	0,30	0,02					1,00	0,90	0,45
0,80	<0,005	0,19	0,65	0,16	<0,005					1,00	0,80	0,42
0,90	<0,005	0,31	0,62	0,07						1,00	0,68	0,38
1,00	0,01	0,44	0,53	0,02						0,99	0,56	0,34
Continuous observation along the profile	<0,005											0,54

Table 47

$$b'=0,5 \qquad -30° \leqslant \Theta \leqslant 30° \qquad d=0,6$$

h in units of d	P_0	P_1	P_2	P_3	P_4	P_5	P_6	P_7	$P_{\geqslant 8}$	$P_{\geqslant 1}$	$P_{\geqslant 2}$	Probability of interception by two profiles
0,05									1,00	1,00	1,00	0,61
0,10									1,00	1,00	1,00	0,61
0,15						0,12	0,30	0,09	0,49	1,00	1,00	0,60
0,20					0,29	0,20	0,28	0,21	0,01	1,00	1,00	0,60
0,25				0,24	0,29	0,34	0,13	<0,005		1,00	1,00	0,59
0,30			0,06	0,42	0,34	0,17	0,01			1,00	1,00	0,58
0,35			0,24	0,42	0,31	0,02				1,00	1,00	0,57
0,40			0,42	0,44	0,14	<0,005				1,00	1,00	0,55
0,45		0,05	0,53	0,36	0,06					1,00	0,95	0,53
0,50		0,15	0,54	0,29	0,02					1,00	0,85	0,52
0,60		0,32	0,54	0,14	<0,005					1,00	0,68	0,47
0,70	<0,005	0,48	0,48	0,04						1,00	0,52	0,41
0,80	0,01	0,62	0,37	<0,005						0,99	0,37	0,34
0,90	0,07	0,66	0,28							0,93	0,28	0,28
1,00	0,15	0,61	0,24							0,85	0,24	0,24
Continuous observation along the profile	—											0,61

Table 48

$$b'=0,5 \qquad -90° \leqslant \Theta \leqslant 90° \qquad d=0,6$$

h in units of d	P_0	P_1	P_2	P_3	P_4	P_5	P_6	P_7	$P_{\geqslant 8}$	$P_{\geqslant 1}$	$P_{\geqslant 2}$	Probability of interception by two profiles
0,05	0,03	<0,005	<0,005	<0,005	<0,005	<0,005	<0,005	<0,005	0,97	0,97	0,97	0,31
0,10	0,03	<0,005	<0,005	<0,005	<0,005	0,01	0,01	0,01	0,94	0,97	0,97	0,31
0,15	0,03	<0,005	<0,005	0,01	0,01	0,06	0,22	0,18	0,49	0,97	0,97	0,31
0,20	0,03	<0,005	0,01	0,02	0,16	0,27	0,26	0,19	0,06	0,97	0,97	0,30
0,25	0,03	0,01	0,02	0,14	0,33	0,30	0,16	0,02		0,97	0,97	0,30
0,30	0,03	0,01	0,05	0,35	0,36	0,19	0,02			0,97	0,96	0,29
0,35	0,03	0,01	0,17	0,44	0,30	0,05				0,97	0,96	0,29
0,40	0,03	0,02	0,32	0,46	0,17	<0,005				0,97	0,95	0,28
0,45	0,03	0,05	0,46	0,40	0,07					0,97	0,92	0,27
0,50	0,03	0,10	0,54	0,30	0,02					0,97	0,87	0,26
0,60	0,03	0,25	0,59	0,13	<0,005					0,97	0,72	0,24
0,70	0,04	0,40	0,52	0,04						0,96	0,56	0,21
0,80	0,05	0,55	0,40	<0,005						0,96	0,40	0,17
0,90	0,07	0,65	0,28							0,93	0,28	0,14
1,00	0,11	0,69	0,20							0,89	0,20	0,11
Continuous observation along the profile	0,03											0,31

Table 49

b′=0,5   —30° ≤ Θ ≤ 30°   d=0,7

h in units of d	P_0	P_1	P_2	P_3	P_4	P_5	P_6	P_7	$P_{\geqslant 8}$	$P_{\geqslant 1}$	$P_{\geqslant 2}$	Probability of interception by two profiles
0,05									1,00	1,00	1,00	0,38
0,10							0,04	0,47	0,49	1,00	1,00	0,38
0,15					0,16	0,50	0,18	0,15	0,01	1,00	1,00	0,37
0,20				0,28	0,48	0,21	0,04			1,00	1,00	0,36
0,25			0,10	0,63	0,22	0,04	<0,005			1,00	1,00	0,35
0,30			0,43	0,47	0,10	<0,005				1,00	1,00	0,34
0,35		0,02	0,70	0,25	0,03					1,00	0,98	0,33
0,40		0,16	0,68	0,15	0,01					1,00	0,84	0,31
0,45		0,31	0,59	0,09	<0,005					1,00	0,69	0,29
0,50		0,44	0,51	0,05	<0,005					1,00	0,56	0,27
0,60	<0,005	0,66	0,33	<0,005						1,00	0,33	0,21
0,70	0,03	0,80	0,17							0,97	0,17	0,16
0,80	0,12	0,75	0,12							0,88	0,12	0,12
0,90	0,22	0,67	0,11							0,78	0,11	0,11
1,00	0,30	0,61	0,10							0,70	0,10	0,10
Continuous observation along the profile	—											0,38

Table 50

b′=0,5   —90° ≤ Θ ≤ 90°   d=0,7

h in units of d	P_0	P_1	P_2	P_3	P_4	P_5	P_6	P_7	$P_{\geqslant 8}$	$P_{\geqslant 1}$	$P_{\geqslant 2}$	Probability of interception by two profiles
0,05	0,07	<0,005	<0,005	<0,005	<0,005	<0,005	<0,005	<0,005	0,91	0,93	0,93	0,17
0,10	0,07	<0,005	<0,005	0,01	0,01	0,02	0,05	0,23	0,61	0,93	0,93	0,17
0,15	0,07	<0,005	0,01	0,02	0,11	0,31	0,22	0,14	0,11	0,93	0,93	0,17
0,20	0,07	0,01	0,02	0,17	0,37	0,22	0,10	0,04	<0,005	0,93	0,92	0,16
0,25	0,07	0,01	0,09	0,43	0,26	0,11	0,02			0,93	0,91	0,16
0,30	0,07	0,02	0,28	0,43	0,16	0,03				0,93	0,90	0,15
0,35	0,07	0,05	0,48	0,32	0,08	<0,005				0,93	0,88	0,14
0,40	0,08	0,12	0,55	0,23	0,02					0,92	0,80	0,14
0,45	0,08	0,22	0,55	0,15	<0,005					0,92	0,70	0,13
0,50	0,08	0,32	0,52	0,08	<0,005					0,92	0,60	0,11
0,60	0,09	0,50	0,39	0,02						0,91	0,41	0,09
0,70	0,11	0,64	0,25	<0,005						0,89	0,25	0,07
0,80	0,15	0,69	0,16							0,85	0,16	0,05
0,90	0,21	0,69	0,10							0,79	0,10	0,04
1,00	0,27	0,67	0,07							0,74	0,07	0,03
Continuous observation along the profile	0,07											0,17

Table 51

b′=0,5   —30° ⩽ Θ ⩽ 30°   d=0,8

h in units of d	P_0	P_1	P_2	P_3	P_4	P_5	P_6	P_7	$P_{\geqslant 8}$	$P_{\geqslant 1}$	$P_{\geqslant 2}$	Probability of interception by two profiles
0,05									1,00	1,00	1,00	0,21
0,10					<0,005	0,18	0,53	0,25	0,04	1,00	1,00	0,20
0,15			<0,005	0,13	0,66	0,20	0,01			1,00	1,00	0,20
0,20			0,10	0,73	0,16	<0,005				1,00	1,00	0,19
0,25		0,01	0,55	0,43	0,02					1,00	0,99	0,18
0,30		0,08	0,80	0,12	<0,005					1,00	0,92	0,16
0,35	<0,005	0,28	0,69	0,03	<0,005					1,00	0,72	0,15
0,40	<0,005	0,47	0,52	0,01						1,00	0,53	0,13
0,45	<0,005	0,63	0,36	<0,005						1,00	0,37	0,10
0,50	0,01	0,75	0,24							0,99	0,24	0,08
0,60	0,06	0,86	0,08							0,94	0,08	0,06
0,70	0,17	0,79	0,04							0,83	0,04	0,04
0,80	0,27	0,69	0,04							0,73	0,04	0,04
0,90	0,35	0,61	0,03							0,65	0,03	0,03
1,00	0,42	0,55	0,03							0,58	0,03	0,03
Continuous observation along the profile	—											0,21

Table 52

b′=0,5   —90° ⩽ Θ ⩽ 90°   d=0,8

h in units of d	P_0	P_1	P_2	P_3	P_4	P_5	P_6	P_7	$P_{\geqslant 8}$	$P_{\geqslant 1}$	$P_{\geqslant 2}$	Probability of interception by two profiles
0,05	0,12	<0,005	<0,005	<0,005	<0,005	0,01	0,01	0,01	0,85	0,88	0,88	0,08
0,10	0,12	<0,005	0,01	0,02	0,03	0,11	0,25	0,19	0,27	0,88	0,88	0,08
0,15	0,12	0,01	0,03	0,10	0,33	0,21	0,09	0,06	0,04	0,88	0,87	0,07
0,20	0,12	0,02	0,10	0,40	0,22	0,09	0,05	<0,005		0,88	0,86	0,07
0,25	0,12	0,04	0,31	0,36	0,12	0,04	<0,005			0,88	0,84	0,07
0,30	0,12	0,09	0,48	0,23	0,07	<0,005				0,88	0,79	0,06
0,35	0,13	0,19	0,51	0,16	0,02					0,87	0,68	0,05
0,40	0,13	0,31	0,46	0,10	<0,005					0,87	0,56	0,05
0,45	0,14	0,41	0,40	0,05						0,86	0,45	0,04
0,50	0,15	0,50	0,33	0,02						0,85	0,35	0,03
0,60	0,18	0,63	0,20	<0,005						0,82	0,20	0,02
0,70	0,23	0,66	0,11							0,77	0,11	0,02
0,80	0,29	0,64	0,06							0,71	0,06	0,01
0,90	0,35	0,61	0,04							0,65	0,04	0,01
1,00	0,41	0,57	0,02							0,59	0,02	0,01
Continuous observation along the profile	0,12											0,08

Table 53

b′=0,5   **−30° ⩽ Θ ⩽ 30°**   **d=0,9**

h in units of d	P_0	P_1	P_2	P_3	P_4	P_5	P_6	P_7	$P_{\geqslant 8}$	$P_{\geqslant 1}$	$P_{\geqslant 2}$	Probability of interception by two profiles
0,05	<0,005	<0,005	<0,005	<0,005	0,01	0,01	0,03	0,08	0,87	1,00	1,00	0,07
0,10	<0,005	<0,005	0,01	0,08	0,22	0,41	0,27	<0,005		1,00	1,00	0,07
0,15	<0,005	0,01	0,13	0,47	0,39	<0,005				1,00	0,99	0,06
0,20	<0,005	0,06	0,46	0,48	<0,005					1,00	0,94	0,05
0,25	<0,005	0,17	0,72	0,11						1,00	0,83	0,04
0,30	0,01	0,37	0,62	<0,005						0,99	0,63	0,03
0,35	0,02	0,58	0,41							0,98	0,41	0,02
0,40	0,04	0,71	0,25							0,96	0,25	0,02
0,45	0,06	0,80	0,14							0,94	0,14	0,01
0,50	0,10	0,84	0,07							0,90	0,07	0,01
0,60	0,20	0,79	0,01							0,80	0,01	0,01
0,70	0,32	0,68	0,01							0,68	0,01	0,01
0,80	0,40	0,59	0,01							0,60	0,01	0,01
0,90	0,47	0,53	0,01							0,53	0,01	0,01
1,00	0,52	0,47	0,01							0,48	0,01	0,01
Continuous observation along the profile	—											0,07

Table 54

b′=0,5   **−90° ⩽ Θ ⩽ 90°**   **d=0,9**

h in units of d	P_0	P_1	P_2	P_3	P_4	P_5	P_6	P_7	$P_{\geqslant 8}$	$P_{\geqslant 1}$	$P_{\geqslant 2}$	Probability of interception by two profiles
0,05	0,17	<0,005	<0,005	0,01	0,01	0,02	0,02	0,04	0,73	0,83	0,83	0,02
0,10	0,17	0,01	0,02	0,05	0,11	0,20	0,19	0,09	0,16	0,83	0,82	0,02
0,15	0,17	0,02	0,09	0,23	0,27	0,11	0,07	0,04	<0,005	0,83	0,81	0,02
0,20	0,18	0,05	0,24	0,33	0,13	0,07	0,01			0,83	0,77	0,02
0,25	0,18	0,11	0,40	0,22	0,08	0,01				0,82	0,71	0,01
0,30	0,18	0,21	0,44	0,14	0,02					0,82	0,61	0,01
0,35	0,19	0,32	0,39	0,09	<0,005					0,81	0,48	0,01
0,40	0,20	0,42	0,33	0,04						0,80	0,37	0,01
0,45	0,22	0,50	0,26	0,02						0,78	0,28	<0,005
0,50	0,24	0,56	0,20	<0,005						0,76	0,20	<0,005
0,60	0,29	0,61	0,10							0,71	0,10	<0,005
0,70	0,36	0,59	0,05							0,64	0,05	<0,005
0,80	0,42	0,56	0,02							0,58	0,02	<0,005
0,90	0,47	0,52	0,01							0,53	0,01	<0,005
1,00	0,52	0,48	0,01							0,48	0,01	<0,005
Continuous observation along the profile	0,17											0,02

Table   55

$$b'=0{,}5 \qquad -30° \leqslant \Theta \leqslant 30° \qquad d=1{,}0$$

h in units of d	P_0	P_1	P_2	P_3	P_4	P_5	P_6	P_7	$P_{\geqslant 8}$	$P_{\geqslant 1}$	$P_{\geqslant 2}$	Probability of interception by two profiles
0,05	0,035	0,009	0,019	0,028	0,039	0,052	0,067	0,087	0,663	0,965	0,956	
0,10	0,040	0,037	0,080	0,137	0,251	0,419	0,037			0,961	0,923	
0,15	0,047	0,087	0,212	0,507	0,146					0,953	0,865	
0,20	0,058	0,167	0,529	0,246						0,942	0,775	
0,25	0,073	0,299	0,614	0,015						0,927	0,629	
0,30	0,091	0,509	0,400							0,909	0,400	
0,35	0,114	0,651	0,235							0,886	0,235	
0,40	0,142	0,735	0,123							0,858	0,123	
0,45	0,176	0,775	0,049							0,824	0,049	
0,50	0,222	0,771	0,007							0,778	0,007	
0,60	0,346	0,654								0,654		
0,70	0,439	0,561								0,561		
0,80	0,509	0,491								0,491		
0,90	0,564	0,436								0,436		
1,00	0,607	0,393								0,393		
Continuous observation along the profile	0,034											

Table   56

$$b'=0{,}5 \qquad -90° \leqslant \Theta \leqslant 90° \qquad d=1{,}0$$

h in units of d	P_0	P_1	P_2	P_3	P_4	P_5	P_6	P_7	$P_{\geqslant 8}$	$P_{\geqslant 1}$	$P_{\geqslant 2}$	Probability of interception by two profiles
0,05	0,230	0,005	0,011	0,016	0,023	0,030	0,038	0,048	0,600	0,770	0,765	
0,10	0,233	0,022	0,046	0,076	0,130	0,220	0,113	0,065	0,096	0,767	0,746	
0,15	0,237	0,050	0,118	0,263	0,189	0,082	0,051	0,010		0,763	0,713	
0,20	0,243	0,095	0,278	0,255	0,095	0,033				0,757	0,662	
0,25	0,252	0,164	0,388	0,153	0,043					0,748	0,584	
0,30	0,262	0,274	0,361	0,097	0,005					0,738	0,464	
0,35	0,275	0,374	0,304	0,046						0,725	0,351	
0,40	0,291	0,453	0,239	0,016						0,709	0,256	
0,45	0,310	0,511	0,176	0,003						0,690	0,179	
0,50	0,334	0,547	0,119							0,666	0,119	
0,60	0,399	0,547	0,054							0,601	0,054	
0,70	0,462	0,515	0,023							0,538	0,023	
0,80	0,517	0,474	0,008							0,483	0,008	
0,90	0,565	0,433	0,002							0,435	0,002	
1,00	0,607	0,393								0,393		
Continuous observation along the profile	0,229											

Table 57

b′ = 0,5 —30° ≤ Θ ≤ 30° d = 1,1

h in units of d	P_0	P_1	P_2	P_3	P_4	P_5	P_6	P_7	$P_{\geqslant 8}$	$P_{\geqslant 1}$	$P_{\geqslant 2}$	Probability of interception by two profiles
0,05	0,123	0,010	0,020	0,032	0,044	0,059	0,078	0,107	0,526	0,877	0,867	
0,10	0,128	0,041	0,090	0,161	0,357	0,222	<0,0005			0,872	0,831	
0,15	0,136	0,098	0,258	4,481	0,027					0,864	0,766	
0,20	0,149	0,191	0,549	0,111						0,851	0,660	
0,25	0,165	0,372	0,463	<0,0005						0,835	0,463	
0,30	0,185	0,547	0,267							0,815	0,267	
0,35	0,211	0,650	0,139							0,789	0,139	
0,40	0,244	0,700	0,056							0,756	0,056	
0,45	0,288	0,703	0,009							0,712	0,009	
0,50	0,351	0,649	<0,0005							0,649	<0,0005	
0,60	0,459	0,541								0,541		
0,70	0,536	0,464								0,464		
0,80	0,594	0,406								0,406		
0,90	0,639	0,361								0,361		
1,00	0,675	0,325								0,325		
Continuous observation along the profile	0,121											

Table 58

b′ = 0,5 —90° ≤ Θ ≤ 90° d = 1,1

h in units of d	P_0	P_1	P_2	P_3	P_4	P_5	P_6	P_7	$P_{\geqslant 8}$	$P_{\geqslant 1}$	$P_{\geqslant 2}$	Probability of interception by two profiles
0,05	0,300	0,006	0,012	0,018	0,025	0,033	0,043	0,057	0,504	0,700	0,694	
0,10	0,303	0,024	0,051	0,089	0,176	0,173	0,078	0,050	0,055	0,697	0,673	
0,15	0,308	0,056	0,140	0,276	0,128	0,065	0,027	<0,0005		0,692	0,636	
0,20	0,315	0,108	0,307	0,190	0,071	0,009				0,685	0,577	
0,25	0,324	0,199	0,348	0,113	0,016					0,676	0,477	
0,30	0,336	0,306	0,298	0,060	<0,0005					0,664	0,358	
0,35	0,351	0,393	0,236	0,021						0,649	0,257	
0,40	0,369	0,456	0,171	0,005						0,631	0,176	
0,45	0,392	0,495	0,113	<0,0005						0,608	0,113	
0,50	0,424	0,504	0,073							0,577	0,073	
0,60	0,489	0,481	0,030							0,511	0,030	
0,70	0,547	0,443	0,011							0,453	0,011	
0,80	0,597	0,401	0,002							0,403	0,002	
0,90	0,639	0,361	<0,0005							0,361	<0,0005	
1,00	0,675	0,325								0,325		
Continuous observation along the profile	0,299											

Table 59

b'=0,5   —30° ≤ Θ ≤ 30°   d=1,2

h in units of d	P_0	P_1	P_2	P_3	P_4	P_5	P_6	P_7	$P_{\geq 8}$	$P_{>1}$	$P_{\geq 2}$	Probability of interception by two profiles
0,05	0,196	0,011	0,022	0,035	0,049	0,067	0,092	0,137	0,389	0,804	0,793	
0,10	0,202	0,045	0,101	0,194	0,392	0,066				0,798	0,753	
0,15	0,211	0,109	0,331	0,348	0,001					0,789	0,680	
0,20	0,225	0,220	0,522	0,033						0,776	0,555	
0,25	0,242	0,424	0,333							0,758	0,333	
0,30	0,266	0,560	0,175							0,734	0,175	
0,35	0,295	0,630	0,075							0,705	0,075	
0,40	0,335	0,649	0,017							0,665	0,017	
0,45	0,394	0,606	<0,0005							0,606	<0,0005	
0,50	0,455	0,545								0,545		
0,60	0,545	0,455								0,455		
0,70	0,610	0,390								0,390		
0,80	0,659	0,341								0,341		
0,90	0,697	0,303								0,303		
1,00	0,727	0,273								0,273		
Continuous observation along the profile	0,195											

Table 60

b'=0,5   —90° ≤ Θ ≤ 90°   d=1,2

h in units of d	P_0	P_1	P_2	P_3	P_4	P_5	P_6	P_7	$P_{\geq 8}$	$P_{>1}$	$P_{\geq 2}$	Probability of interception by two profiles
0,05	0,359	0,006	0,013	0,020	0,028	0,038	0,050	0,071	0,415	0,641	0,635	
0,10	0,362	0,026	0,057	0,105	0,203	0,120	0,060	0,041	0,027	0,638	0,612	
0,15	0,367	0,062	0,172	0,246	0,095	0,050	0,007			0,633	0,571	
0,20	0,375	0,123	0,315	0,141	0,046	0,001				0,625	0,503	
0,25	0,385	0,228	0,301	0,081	0,004					0,615	0,386	
0,30	0,398	0,326	0,243	0,032						0,602	0,275	
0,35	0,415	0,399	0,178	0,008						0,585	0,186	
0,40	0,436	0,446	0,117	<0,0005						0,564	0,118	
0,45	0,466	0,462	0,072							0,534	0,072	
0,50	0,499	0,456	0,045							0,501	0,045	
0,60	0,562	0,422	0,016							0,439	0,016	
0,70	0,614	0,381	0,004							0,386	0,004	
0,80	0,659	0,340	<0,0005							0,341	<0,0005	
0,90	0,697	0,303								0,303		
1,00	0,727	0,273								0,273		
Continuous observation along the profile	0,358											

Table 61

$$b'=0,5 \qquad -30° \leqslant \Theta \leqslant 30° \qquad d=1,3$$

h in units of d	P_0	P_1	P_2	P_3	P_4	P_5	P_6	P_7	$P_{\geqslant 8}$	$P_{\geqslant 1}$	$P_{\geqslant 2}$	Probability of interception by two profiles
0,05	0,258	0,012	0,024	0,038	0,055	0,077	0,111	0,204	0,221	0,742	0,730	
0,10	0,264	0,050	0,112	0,251	0,316	0,006				0,736	0,686	
0,15	0,275	0,121	0,386	0,219						0,726	0,605	
0,20	0,289	0,263	0,445	0,003						0,711	0,448	
0,25	0,309	0,453	0,238							0,691	0,238	
0,30	0,335	0,556	0,110							0,665	0,110	
0,35	0,369	0,597	0,033							0,631	0,033	
0,40	0,421	0,578	0,002							0,579	0,002	
0,45	0,484	0,516								0,516		
0,50	0,535	0,465								0,465		
0,60	0,613	0,387								0,387		
0,70	0,668	0,332								0,332		
0,80	0,710	0,290								0,290		
0,90	0,742	0,258								0,258		
1,00	0,768	0,232								0,232		
Continuous observation along the profile	0,256											

Table 62

$$b'=0,5 \qquad -90° \leqslant \Theta \leqslant 90° \qquad d=1,3$$

h in units of d	P_0	P_1	P_2	P_3	P_4	P_5	P_6	P_7	$P_{\geqslant 8}$	$P_{\geqslant 1}$	$P_{\geqslant 2}$	Probability of interception by two profiles
0,05	0,408	0,007	0,014	0,022	0,031	0,043	0,059	0,097	0,319	0,592	0,585	
0,10	0,412	0,029	0,063	0,129	0,193	0,085	0,049	0,033	0,007	0,588	0,560	
0,15	0,417	0,069	0,201	0,205	0,076	0,031	<0,0005			0,583	0,514	
0,20	0,426	0,142	0,300	0,108	0,024					0,574	0,432	
0,25	0,437	0,251	0,257	0,055	<0,0005					0,563	0,312	
0,30	0,452	0,338	0,194	0,016						0,548	0,211	
0,35	0,471	0,397	0,130	0,002						0,529	0,133	
0,40	0,497	0,425	0,078							0,503	0,078	
0,45	0,530	0,423	0,047							0,470	0,047	
0,50	0,563	0,410	0,028							0,437	0,028	
0,60	0,621	0,371	0,008							0,379	0,008	
0,70	0,669	0,330	0,001							0,331	0,001	
0,80	0,710	0,290								0,280		
0,90	0,742	0,258								0,258		
1,00	0,768	0,232								0,232		
Continuous observation along the profile	0,407											

Table 63

$b'=0,5$ $-30° \leqslant \Theta \leqslant 30°$ $d=1,4$

h in units of d	P_0	P_1	P_2	P_3	P_4	P_5	P_6	P_7	$P_{\geqslant 8}$	$P_{\geqslant 1}$	$P_{\geqslant 2}$	Probability of interception by two profiles
0,05	0,312	0,013	0,026	0,042	0,061	0,088	0,139	0,255	0,064	0,688	0,675	
0,10	0,318	0,054	0,126	0,310	0,192					0,682	0,628	
0,15	0,329	0,134	0,409	0,128						0,671	0,537	
0,20	0,345	0,308	0,347							0,655	0,347	
0,25	0,367	0,465	0,168							0,633	0,168	
0,30	0,396	0,540	0,064							0,604	0,064	
0,35	0,437	0,554	0,009							0,563	0,009	
0,40	0,499	0,501								0,501		
0,45	0,555	0,445								0,445		
0,50	0,599	0,401								0,401		
0,60	0,666	0,334								0,334		
0,70	0,714	0,286								0,286		
0,80	0,750	0,250								0,250		
0,90	0,777	0,223								0,223		
1,00	0,800	0,200								0,200		
Continuous observation along the profile	0,310											

Table 64

$b'=0,5$ $-90° \leqslant \Theta \leqslant 90°$ $d=1,4$

h in units of d	P_0	P_1	P_2	P_3	P_4	P_5	P_6	P_7	$P_{\geqslant 8}$	$P_{\geqslant 1}$	$P_{\geqslant 2}$	Probability of interception by two profiles
0,05	0,451	0,007	0,015	0,024	0,034	0,048	0,072	0,121	0,228	0,549	0,542	
0,10	0,454	0,031	0,070	0,156	0,160	0,067	0,041	0,020	$<0,0005$	0,546	0,515	
0,15	0,461	0,076	0,221	0,166	0,063	0,014				0,539	0,464	
0,20	0,470	0,164	0,271	0,085	0,010					0,530	0,366	
0,25	0,482	0,267	0,218	0,033						0,518	0,251	
0,30	0,498	0,342	0,152	0,007						0,502	0,159	
0,35	0,520	0,387	0,093	$<0,0005$						0,480	0,093	
0,40	0,552	0,395	0,053							0,448	0,053	
0,45	0,585	0,385	0,030							0,415	0,030	
0,50	0,616	0,368	0,017							0,384	0,017	
0,60	0,670	0,327	0,003							0,331	0,003	
0,70	0,714	0,286	$<0,0005$							0,286	$<0,0005$	
0,80	0,750	0,250								0,250		
0,90	0,777	0,223								0,223		
1,00	0,800	0,200								0,200		
Continuous observation along the profile	0,449											

$$b' = 0,5 \qquad -30° \leqslant \Theta \leqslant 30° \qquad d = 1,5$$

h in units of d	P_0	P_1	P_2	P_3	P_4	P_5	P_6	P_7	$P_{\geqslant 8}$	$P_{>1}$	$P_{\geqslant 2}$	Probability of interception by two profiles
0,05	0,358	0,014	0,029	0,046	0,068	0,102	0,195	0,183	0,006	0,642	0,628	
0,10	0,365	0,058	0,141	0,338	0,098					0,635	0,577	
0,15	0,377	0,148	0,410	0,065						0,624	0,475	
0,20	0,394	0,339	0,267							0,606	0,267	
0,25	0,418	0,466	0,116							0,582	0,116	
0,30	0,451	0,517	0,033							0,549	0,033	
0,35	0,502	0,497	0,001							0,498	0,001	
0,40	0,564	0,436								0,436		
0,45	0,612	0,388								0,388		
0,50	0,651	0,349								0,349		
0,60	0,709	0,291								0,291		
0,70	0,751	0,249								0,249		
0,80	0,782	0,218								0,218		
0,90	0,806	0,194								0,194		
1,00	0,825	0,175								0,175		
Continuous observation along the profile	0,356											

Table 66

$$b' = 0,5 \qquad -90° \leqslant \Theta \leqslant 90° \qquad d = 1,5$$

h in units of d	P_0	P_1	P_2	P_3	P_4	P_5	P_6	P_7	$P_{\geqslant 8}$	$P_{>1}$	$P_{\geqslant 2}$	Probability of interception by two profiles
0,05	0,487	0,008	0,017	0,026	0,038	0,055	0,094	0,108	0,168	0,513	0,505	
0,10	0,491	0,034	0,078	0,175	0,126	0,055	0,034	0,007		0,509	0,475	
0,15	0,498	0,083	0,233	0,132	0,049	0,005				0,502	0,419	
0,20	0,508	0,183	0,241	0,065	0,003					0,492	0,309	
0,25	0,522	0,278	0,182	0,019						0,478	0,201	
0,30	0,540	0,341	0,117	0,002						0,460	0,119	
0,35	0,566	0,369	0,065							0,434	0,065	
0,40	0,600	0,365	0,036							0,401	0,036	
0,45	0,631	0,349	0,019							0,369	0,019	
0,50	0,660	0,330	0,010							0,340	0,010	
0,60	0,710	0,289	0,001							0,290	0,001	
0,70	0,751	0,249								0,249		
0,80	0,782	0,218								0,218		
0,90	0,806	0,194								0,194		
1,00	0,825	0,175								0,175		
Continuous observation along the profile	0,486											

Table 67

$$b'=0,5 \qquad -30° \leqslant \Theta \leqslant 30° \qquad d=1,7$$

h in units of d	P_0	P_1	P_2	P_3	P_4	P_5	P_6	P_7	$P_{\geqslant 8}$	$P_{\geqslant 1}$	$P_{\geqslant 2}$	Probability of interception by two profiles
0,05	0,434	0,016	0,033	0,054	0,084	0,153	0,212	0,014		0,566	0,550	
0,10	0,442	0,068	0,187	0,296	0,007					0,558	0,491	
0,15	0,455	0,188	0,352	0,005						0,545	0,357	
0,20	0,476	0,369	0,155							0,524	0,155	
0,25	0,505	0,447	0,048							0,495	0,048	
0,30	0,549	0,448	0,002							0,451	0,002	
0,35	0,612	0,388								0,388		
0,40	0,660	0,340								0,340		
0,45	0,698	0,302								0,302		
0,50	0,728	0,272								0,272		
0,60	0,774	0,226								0,226		
0,70	0,806	0,194								0,194		
0,80	0,830	0,170								0,170		
0,90	0,849	0,151								0,151		
1,00	0,864	0,136								0,136		
Continuous observation along the profile	0,431											

Table 68

$$b'=0,5 \qquad -90° \leqslant \Theta \leqslant 90° \qquad d=1,7$$

h in units of d	P_0	P_1	P_2	P_3	P_4	P_5	P_6	P_7	$P_{\geqslant 8}$	$P_{\geqslant 1}$	$P_{\geqslant 2}$	Probability of interception by two profiles
0,05	0,548	0,009	0,019	0,031	0,046	0,077	0,113	0,057	0,100	0,452	0,443	
0,10	0,553	0,039	0,100	0,180	0,076	0,040	0,012			0,448	0,409	
0,15	0,560	0,103	0,229	0,086	0,022					0,440	0,337	
0,20	0,572	0,209	0,186	0,033						0,428	0,219	
0,25	0,588	0,285	0,122	0,005						0,412	0,127	
0,30	0,612	0,324	0,065							0,388	0,065	
0,35	0,645	0,322	0,033							0,355	0,033	
0,40	0,677	0,307	0,016							0,324	0,016	
0,45	0,705	0,288	0,007							0,295	0,007	
0,50	0,731	0,267	0,002							0,269	0,002	
0,60	0,774	0,226								0,226		
0,70	0,806	0,194								0,194		
0,80	0,830	0,170								0,170		
0,90	0,849	0,151								0,151		
1,00	0,864	0,136								0,136		
Continuous observation along the profile	0,546											

Table 69

$$b'=0,5 \qquad -30° \leqslant \Theta \leqslant 30° \qquad d=2,0$$

h in units of d	P_0	P_1	P_2	P_3	P_4	P_5	P_6	P_7	$P_{\geqslant 8}$	$P_{\geqslant 1}$	$P_{\geqslant 2}$	Probability of interception by two profiles
0,05	0,520	0,019	0,040	0,068	0,125	0,210	0,018			0,480	0,462	
0,10	0,529	0,083	0,264	0,123						0,471	0,388	
0,15	0,545	0,255	0,200							0,455	0,200	
0,20	0,571	0,368	0,062							0,429	0,062	
0,25	0,611	0,385	0,004							0,389	0,004	
0,30	0,673	0,327								0,327		
0,35	0,720	0,280								0,280		
0,40	0,755	0,245								0,245		
0,45	0,782	0,218								0,218		
0,50	0,804	0,196								0,196		
0,60	0,836	0,164								0,164		
0,70	0,860	0,140								0,140		
0,80	0,877	0,123								0,123		
0,90	0,891	0,109								0,109		
1,00	0,902	0,098								0,098		
Continuous observation along the profile	0,517											

Table 70

$$b'=0,5 \qquad -90° \leqslant \Theta \leqslant 90° \qquad d=2,0$$

h in units of d	P_0	P_1	P_2	P_3	P_4	P_5	P_6	P_7	$P_{\geqslant 8}$	$P_{\geqslant 1}$	$P_{\geqslant 2}$	Probability of interception by two profiles
0,05	0,616	0,011	0,023	0,038	0,065	0,110	0,057	0,032	0,048	0,384	0,373	
0,10	0,622	0,047	0,139	0,128	0,048	0,016	<0,0005			0,378	0,331	
0,15	0,631	0,137	0,181	0,049	0,003					0,369	0,232	
0,20	0,645	0,227	0,120	0,008						0,355	0,128	
0,25	0,667	0,273	0,060							0,333	0,060	
0,30	0,700	0,273	0,027							0,300	0,027	
0,35	0,731	0,257	0,012							0,269	0,012	
0,40	0,759	0,237	0,004							0,241	0,004	
0,45	0,783	0,216	0,001							0,217	0,001	
0,50	0,804	0,196								0,196		
0,60	0,836	0,164								0,164		
0,70	0,860	0,140								0,140		
0,80	0,877	0,123								0,123		
0,90	0,891	0,109								0,109		
1,00	0,902	0,098								0,098		
Continuous observation along the profile	0,615											

Table 71

$$b'=0{,}3 \qquad -30' \leqslant \Theta \leqslant 30° \qquad d=0{,}5$$

h in units of d	P_0	P_1	P_2	P_3	P_4	P_5	P_6	P_7	$P_{\geqslant 8}$	$P_{\geqslant 1}$	$P_{\geqslant 2}$	Probability of interception by two profiles
0,05									1,00	1,00	1,00	0,92
0,10						<0,005	0,05	0,08	0,87	1,00	1,00	0,91
0,15				<0,005	0,07	0,13	0,32	0,40	0,08	1,00	1,00	0,90
0,20			<0,005	0,09	0,26	0,49	0,16	<0,005		1,00	1,00	0,88
0,25			0,06	0,27	0,50	0,16	<0,005			1,00	1,00	0,87
0,30		<0,005	0,16	0,55	0,29	0,01				1,00	1,00	0,84
0,35		0,04	0,35	0,50	0,11					1,00	0,96	0,81
0,40		0,09	0,50	0,36	0,04					1,00	0,91	0,78
0,45		0,16	0,60	0,23	0,01					1,00	0,84	0,74
0,50		0,24	0,64	0,12	<0,005					1,00	0,76	0,69
0,60	0,01	0,43	0,56	0,01						0,99	0,57	0,56
0,70	0,06	0,53	0,41							0,94	0,41	0,41
0,80	0,13	0,56	0,31							0,87	0,31	0,31
0,90	0,20	0,55	0,25							0,80	0,25	0,25
1,00	0,27	0,51	0,21							0,73	0,21	0,21
Continuous observation along the profile	—											0,92

Table 72

$$b'=0{,}3 \qquad -90° \leqslant \Theta \leqslant 90° \qquad d=0{,}5$$

h in units of d	P_0	P_1	P_2	P_3	P_4	P_5	P_6	P_7	$P_{\geqslant 8}$	$P_{\geqslant 1}$	$P_{\geqslant 2}$	Probability of interception by two profiles
0,05	0,07	<0,005	<0,005	<0,005	<0,005	<0,005	<0,005	<0,005	0,93	0,93	0,93	0,46
0,10	0,07	<0,005	<0,005	<0,005	<0,005	0,01	0,03	0,08	0,81	0,93	0,93	0,46
0,15	0,07	<0,005	<0,005	0,01	0,05	0,15	0,23	0,26	0,23	0,93	0,93	0,46
0,20	0,07	<0,005	0,01	0,07	0,25	0,32	0,17	0,04	0,06	0,93	0,93	0,45
0,25	0,07	0,01	0,04	0,25	0,37	0,17	0,05	0,03	0,01	0,93	0,92	0,44
0,30	0,07	0,01	0,14	0,42	0,26	0,06	0,03	0,01		0,93	0,92	0,43
0,35	0,07	0,03	0,29	0,42	0,14	0,04	0,01			0,93	0,90	0,41
0,40	0,07	0,07	0,43	0,33	0,08	0,02				0,93	0,86	0,40
0,45	0,07	0,13	0,50	0,24	0,05	<0,005				0,93	0,80	0,38
0,50	0,07	0,20	0,53	0,17	0,03					0,93	0,73	0,35
0,60	0,08	0,35	0,49	0,07	<0,005					0,92	0,57	0,29
0,70	0,10	0,48	0,39	0,03						0,90	0,42	0,22
0,80	0,14	0,55	0,30	0,01						0,86	0,31	0,17
0,90	0,19	0,57	0,23	<0,005						0,81	0,24	0,14
1,00	0,24	0,58	0,18							0,76	0,18	0,11
Continuous observation along the profile	0,07											0,046

48

Table 73

$$b'=0,3 \qquad -30° \leqslant \Theta \leqslant 30° \qquad d=0,6$$

h in units of d	P_0	P_1	P_2	P_3	P_4	P_5	P_6	P_7	$P_{\geqslant 8}$	$P_{\geqslant 1}$	$P_{\geqslant 2}$	Probability of interception by two profiles
0,05									1,00	1,00	1,00	0,60
0,10					<0,005	0,33	0,15	0,20	0,32	1,00	1,00	0,59
0,15				0,23	0,31	0,33	0,13	<0,005		1,00	1,00	0,58
0,20			0,18	0,42	0,33	0,06	<0,005			1,00	1,00	0,56
0,25		<0,005	0,48	0,41	0,11					1,00	1,00	0,53
0,30		0,14	0,56	0,28	0,02					1,00	0,86	0,50
0,35		0,29	0,56	0,15	<0,005					1,00	0,71	0,47
0,40		0,42	0,53	0,05	<0,005					1,00	0,58	0,42
0,45	<0,005	0,55	0,44	0,01						1,00	0,45	0,37
0,50	0,01	0,66	0,32	<0,005						0,99	0,32	0,31
0,60	0,12	0,68	0,21							0,88	0,21	0,21
0,70	0,21	0,65	0,14							0,79	0,14	0,14
0,80	0,29	0,60	0,11							0,71	0,11	0,11
0,90	0,37	0,54	0,09							0,63	0,09	0,09
1,00	0,43	0,49	0,08							0,57	0,08	0,08
Continuous observation along the profile	—											0,60

Table 74

$$b'=0,3 \qquad -90° \leqslant \Theta \leqslant 90° \qquad d=0,6$$

h in units of d	P_0	P_1	P_2	P_3	P_4	P_5	P_6	P_7	$P_{\geqslant 8}$	$P_{\geqslant 1}$	$P_{\geqslant 2}$	Probability of interception by two profiles
0,05	0,11	<0,005	<0,005	<0,005	<0,005	<0,005	<0,005	<0,005	0,87	0,89	0,89	0,27
0,10	0,11	<0,005	<0,005	0,01	0,01	0,16	0,18	0,17	0,35	0,89	0,89	0,27
0,15	0,11	0,01	0,01	0,12	0,27	0,25	0,12	0,04	0,07	0,89	0,88	0,26
0,20	0,11	0,01	0,10	0,35	0,26	0,09	0,04	0,03	0,02	0,89	0,88	0,25
0,25	0,11	0,02	0,32	0,36	0,12	0,04	0,03	0,01		0,89	0,87	0,24
0,30	0,11	0,09	0,45	0,25	0,07	0,03	<0,005			0,89	0,80	0,23
0,35	0,12	0,19	0,48	0,16	0,05	0,01				0,88	0,70	0,21
0,40	0,12	0,30	0,45	0,10	0,03	<0,005				0,88	0,59	0,19
0,45	0,12	0,40	0,40	0,07	0,01					0,88	0,48	0,17
0,50	0,13	0,49	0,33	0,05	<0,005					0,87	0,38	0,14
0,60	0,18	0,57	0,23	0,02						0,82	0,25	0,10
0,70	0,25	0,58	0,17	0,01						0,75	0,18	0,07
0,80	0,31	0,57	0,12	<0,005						0,69	0,12	0,06
0,90	0,36	0,55	0,09							0,64	0,09	0,04
1,00	0,41	0,53	0,06							0,59	0,06	0,03
Continuous observation along the profile	0,11											0,27

Table 75

b′=0,3   −30° ≤ Θ ≤ 30°   d=0,7

h in units of d	P_0	P_1	P_2	P_3	P_4	P_5	P_6	P_7	$P_{\geqslant 8}$	$P_{\geqslant 1}$	$P_{\geqslant 2}$	Probability of interception by two profiles
0,05								<0,005	1,00	1,00	1,00	0,37
0,10				<0,005	0,45	0,32	0,19	0,04	<0,005	1,00	1,00	0,36
0,15			0,10	0,64	0,22	0,04	<0,005			1,00	1,00	0,34
0,20		<0,005	0,63	0,32	0,04					1,00	1,00	0,32
0,25		0,21	0,67	0,12	<0,005					1,00	0,79	0,29
0,30		0,44	0,52	0,04	<0,005					1,00	0,56	0,26
0,35	<0,005	0,63	0,36	0,01						1,00	0,37	0,21
0,40	0,02	0,76	0,22							0,98	0,22	0,17
0,45	0,07	0,79	0,14							0,93	0,14	0,14
0,50	0,15	0,75	0,11							0,85	0,11	0,11
0,60	0,27	0,66	0,07							0,73	0,07	0,07
0,70	0,36	0,59	0,05							0,64	0,05	0,05
0,80	0,44	0,52	0,04							0,56	0,04	0,04
0,90	0,50	0,46	0,04							0,50	0,04	0,04
1,00	0,55	0,42	0,03							0,45	0,03	0,03
Continuous observation along the profile	—											0,37

Table 76

b′=0,3   −90° ≤ Θ ≤ 90°   d=0,7

h in units of d	P_0	P_1	P_2	P_3	P_4	P_5	P_6	P_7	$P_{\geqslant 8}$	$P_{\geqslant 1}$	$P_{\geqslant 2}$	Probability of interception by two profiles
0,05	0,15	<0,005	<0,005	<0,005	0,01	0,01	0,02	0,02	0,79	0,85	0,84	0,15
0,10	0,15	<0,005	0,01	0,03	0,22	0,25	0,15	0,06	0,12	0,85	0,84	0,15
0,15	0,16	0,01	0,08	0,37	0,21	0,08	0,04	0,03	0,04	0,84	0,83	0,14
0,20	0,16	0,03	0,36	0,29	0,09	0,04	0,03	0,01	<0,005	0,84	0,82	0,13
0,25	0,16	0,13	0,46	0,16	0,05	0,03	0,01			0,84	0,71	0,12
0,30	0,16	0,27	0,42	0,10	0,04	0,01				0,84	0,56	0,10
0,35	0,17	0,40	0,34	0,07	0,02	<0,005				0,83	0,43	0,08
0,40	0,18	0,51	0,26	0,05	0,01					0,82	0,31	0,07
0,45	0,20	0,56	0,20	0,04	<0,005					0,80	0,23	0,05
0,50	0,24	0,57	0,16	0,02						0,76	0,18	0,04
0,60	0,32	0,56	0,11	<0,005						0,68	0,12	0,03
0,70	0,39	0,53	0,08	<0,005						0,61	0,08	0,02
0,80	0,45	0,50	0,05							0,55	0,05	0,02
0,90	0,50	0,47	0,03							0,50	0,03	0,01
1,00	0,54	0,44	0,02							0,46	0,02	0,01
Continuous observation along the profile	0,15											0,15

Table 77

$b'=0,3 \qquad -30° \leqslant \Theta \leqslant 30° \qquad d=0,8$

h in units of d	P_0	P_1	P_2	P_3	P_4	P_5	P_6	P_7	$P_{\geqslant 8}$	$P_{\geqslant 1}$	$P_{\geqslant 2}$	Probability of interception by two profiles
0,05					<0,005	0,01	0,15	0,40	0,44	1,00	1,00	0,20
0,10			0,01	0,36	0,58	0,05	<0,005			1,00	1,00	0,19
0,15		0,01	0,55	0,43	0,02					1,00	0,99	0,17
0,20		0,20	0,76	0,04	<0,005					1,00	0,80	0,14
0,25	<0,005	0,53	0,47	<0,005						1,00	0,47	0,11
0,30	0,01	0,75	0,24							0,99	0,24	0,08
0,35	0,05	0,84	0,10							0,95	0,10	0,06
0,40	0,13	0,82	0,05							0,87	0,05	0,05
0,45	0,22	0,74	0,04							0,78	0,04	0,04
0,50	0,29	0,68	0,03							0,71	0,03	0,03
0,60	0,40	0,58	0,02							0,60	0,02	0,02
0,70	0,49	0,50	0,01							0,51	0,01	0,01
0,80	0,55	0,44	0,01							0,45	0,01	0,01
0,90	0,60	0,39	0,01							0,40	0,01	0,01
1,00	0,64	0,35	0,01							0,36	0,01	0,01
Continuous observation along the profile	—											0,20

Table 78

$b'=0,3 \qquad -90° \leqslant \Theta \leqslant 90° \qquad d=0,8$

h in units of d	P_0	P_1	P_2	P_3	P_4	P_5	P_6	P_7	$P_{\geqslant 8}$	$P_{\geqslant 1}$	$P_{\geqslant 2}$	Probability of interception by two profiles
0,05	0,20	<0,005	0,01	0,01	0,01	0,02	0,08	0,17	0,50	0,80	0,80	0,07
0,10	0,20	0,01	0,03	0,18	0,30	0,12	0,05	0,03	0,08	0,80	0,79	0,07
0,15	0,20	0,03	0,27	0,30	0,09	0,04	0,03	0,02	0,01	0,80	0,77	0,06
0,20	0,21	0,13	0,44	0,13	0,05	0,03	0,02	<0,005		0,79	0,66	0,05
0,25	0,21	0,30	0,36	0,08	0,04	0,01	<0,005			0,79	0,49	0,04
0,30	0,22	0,43	0,27	0,06	0,02	<0,005				0,78	0,35	0,03
0,35	0,24	0,52	0,19	0,05	0,01					0,76	0,24	0,02
0,40	0,28	0,55	0,14	0,03	<0,005					0,72	0,17	0,02
0,45	0,33	0,55	0,11	0,02						0,67	0,13	0,01
0,50	0,37	0,53	0,09	0,01						0,63	0,10	0,01
0,60	0,44	0,50	0,06	<0,005						0,56	0,06	0,01
0,70	0,51	0,46	0,03							0,49	0,03	<0,005
0,80	0,56	0,42	0,02							0,44	0,02	<0,005
0,90	0,60	0,39	0,01							0,40	0,01	<0,005
1,00	0,64	0,35	0,01							0,36	0,01	<0,005
Continuous observation along the profile	0,20											0,07

Table 79

$b'=0,3 \qquad -30° \leqslant \Theta \leqslant 30° \qquad d=0,9$

h in units of d	P_0	P_1	P_2	P_3	P_4	P_5	P_6	P_7	$P_{\geqslant 8}$	$P_{\geqslant 1}$	$P_{\geqslant 2}$	Probability of interception by two profiles
0,05	<0,005	<0,005	0,01	0,03	0,10	0,19	0,33	0,31	0,02	1,00	1,00	0,06
0,10	<0,005	0,02	0,21	0,58	0,18					1,00	0,97	0,05
0,15	0,01	0,17	0,70	0,12						0,99	0,83	0,04
0,20	0,02	0,51	0,47							0,98	0,47	0,02
0,25	0,05	0,74	0,21							0,95	0,21	0,01
0,30	0,10	0,83	0,07							0,90	0,07	0,01
0,35	0,18	0,81	0,01							0,82	0,01	0,01
0,40	0,28	0,72	0,01							0,72	0,01	0,01
0,45	0,36	0,64	<0,005							0,64	<0,005	<0,005
0,50	0,42	0,57	<0,005							0,58	<0,005	<0,005
0,60	0,52	0,48	<0,005							0,48	<0,005	<0,005
0,70	0,59	0,41	<0,005							0,41	<0,005	<0,005
0,80	0,64	0,36	<0,005							0,36	<0,005	<0,005
0,90	0,68	0,32	<0,005							0,32	<0,005	<0,005
1,00	0,71	0,29	<0,005							0,29	<0,005	<0,005
Continuous observation along the profile	<0,005											0,07

Table 80

$b'=0,3 \qquad -90° \leqslant \Theta \leqslant 90° \qquad d=0,9$

h in units of d	P_0	P_1	P_2	P_3	P_4	P_5	P_6	P_7	$P_{\geqslant 8}$	$P_{\geqslant 1}$	$P_{\geqslant 2}$	Probability of interception by two profiles
0,05	0,25	0,01	0,01	0,02	0,05	0,09	0,14	0,16	0,27	0,75	0,75	0,02
0,10	0,25	0,03	0,11	0,27	0,17	0,07	0,04	0,02	0,05	0,75	0,72	0,02
0,15	0,25	0,10	0,35	0,18	0,06	0,03	0,02	0,01	<0,005	0,75	0,65	0,01
0,20	0,26	0,25	0,34	0,08	0,04	0,02	<0,005			0,74	0,49	0,01
0,25	0,28	0,39	0,24	0,06	0,03	<0,005				0,72	0,33	<0,005
0,30	0,31	0,48	0,17	0,04	0,01					0,69	0,22	<0,005
0,35	0,34	0,51	0,12	0,03	<0,005					0,66	0,14	<0,005
0,40	0,39	0,50	0,09	0,01						0,61	0,10	<0,005
0,45	0,44	0,49	0,07	0,01						0,56	0,08	<0,005
0,50	0,48	0,47	0,06	<0,005						0,52	0,06	<0,005
0,60	0,55	0,42	0,03							0,45	0,03	<0,005
0,70	0,60	0,38	0,02							0,40	0,02	<0,005
0,80	0,64	0,35	0,01							0,36	0,01	<0,005
0,90	0,68	0,32	<0,005							0,32	<0,005	<0,005
1,00	0,71	0,29	<0,005							0,29	<0,005	<0,005
Continuous observation along the profile	0,25											0,02

Table 81

$$b' = 0{,}3 \qquad -30° \leqslant \Theta \leqslant 30° \qquad d = 1{,}0$$

h in units of d	P_0	P_1	P_2	P_3	P_4	P_5	P_6	P_7	$P_{\geqslant 8}$	$P_{>1}$	$P_{>2}$	Probability of interception by two profiles
0,05	0,045	0,025	0,052	0,085	0,131	0,224	0,377	0,060		0,955	0,930	
0,10	0,058	0,108	0,286	0,519	0,030					0,942	0,835	
0,15	0,079	0,291	0,610	0,020						0,921	0,630	
0,20	0,111	0,599	0,289							0,889	0,289	
0,25	0,157	0,744	0,099							0,843	0,099	
0,30	0,225	0,765	0,010							0,775	0,010	
0,35	0,327	0,673								0,673		
0,40	0,411	0,589								0,589		
0,45	0,477	0,523								0,523		
0,50	0,529	0,471								0,471		
0,60	0,607	0,393								0,393		
0,70	0,663	0,337								0,337		
0,80	0,706	0,294								0,294		
0,90	0,738	0,262								0,262		
1,00	0,764	0,236								0,236		
Continuous observation along the profile	0,041											

Table 82

$$b' = 0{,}3 \qquad -90° \leqslant \Theta \leqslant 90° \qquad d = 1{,}0$$

h in units of d	P_0	P_1	P_2	P_3	P_4	P_5	P_6	P_7	$P_{\geqslant 8}$	$P_{>1}$	$P_{>2}$	Probability of interception by two profiles
0,05	0,304	0,013	0,027	0,043	0,065	0,104	0,170	0,087	0,188	0,696	0,683	
0,10	0,311	0,055	0,138	0,265	0,107	0,047	0,028	0,019	0,030	0,689	0,635	
0,15	0,322	0,143	0,334	0,113	0,043	0,025	0,016	0,003		0,678	0,536	
0,20	0,338	0,298	0,263	0,061	0,029	0,011				0,662	0,364	
0,25	0,361	0,408	0,174	0,044	0,014					0,639	0,232	
0,30	0,393	0,462	0,113	0,030	0,002					0,607	0,145	
0,35	0,441	0,461	0,084	0,015						0,560	0,099	
0,40	0,487	0,443	0,065	0,005						0,513	0,071	
0,45	0,528	0,421	0,050	0,001						0,472	0,051	
0,50	0,565	0,400	0,036							0,435	0,036	
0,60	0,624	0,359	0,017							0,376	0,017	
0,70	0,671	0,322	0,007							0,329	0,007	
0,80	0,708	0,289	0,003							0,292	0,003	
0,90	0,739	0,261	0,001							0,261	0,001	
1,00	0,764	0,236								0,236		
Continuous observation along the profile	0,302											

$b'=0,3$ $\quad -30° \leqslant \Theta \leqslant 30°$ $\quad d=1,1$

h in units of d	P_0	P_1	P_2	P_3	P_4	P_5	P_6	P_7	$P_{\geqslant 8}$	$P_{\geqslant 1}$	$P_{\geqslant 2}$	Probability of interception by two profiles
0,05	0,133	0,028	0,058	0,097	0,159	0,329	0,195	0,002		0,867	0,840	
0,10	0,146	0,121	0,372	0,360	0,001					0,854	0,733	
0,15	0,171	0,361	0,468	0,001						0,830	0,468	
0,20	0,207	0,612	0,181							0,793	0,181	
0,25	0,261	0,700	0,040							0,739	0,040	
0,30	0,351	0,649	<0,0005							0,649	<0,0005	
0,35	0,444	0,556								0,556		
0,40	0,513	0,487								0,487		
0,45	0,567	0,433								0,433		
0,50	0,611	0,389								0,389		
0,60	0,675	0,325								0,325		
0,70	0,722	0,278								0,278		
0,80	0,757	0,243								0,243		
0,90	0,784	0,216								0,216		
1,00	0,805	0,195								0,195		
Continuous observation along the profile	0,128											

$b'=0,3$ $\quad -90° \leqslant \Theta \leqslant 90°$ $\quad d=1,1$

h in units of d	P_0	P_1	P_2	P_3	P_4	P_5	P_6	P_7	$P_{\geqslant 8}$	$P_{\geqslant 1}$	$P_{\geqslant 2}$	Probability of interception by two profiles
0,05	0,368	0,014	0,030	0,049	0,077	0,145	0,125	0,056	0,137	0,632	0,618	
0,10	0,375	0,061	0,174	0,225	0,074	0,036	0,022	0,015	0,018	0,625	0,564	
0,15	0,387	0,173	0,296	0,081	0,034	0,020	0,009	<0,0005		0,613	0,440	
0,20	0,406	0,317	0,205	0,048	0,022	0,003				0,595	0,278	
0,25	0,432	0,402	0,126	0,034	0,005					0,568	0,166	
0,30	0,474	0,423	0,085	0,019	<0,0005					0,526	0,104	
0,35	0,521	0,408	0,064	0,007						0,479	0,071	
0,40	0,564	0,387	0,048	0,002						0,436	0,049	
0,45	0,601	0,365	0,034	<0,0005						0,399	0,034	
0,50	0,633	0,345	0,022							0,367	0,022	
0,60	0,685	0,306	0,009							0,315	0,009	
0,70	0,725	0,271	0,003							0,275	0,003	
0,80	0,757	0,242	0,001							0,243	0,001	
0,90	0,784	0,216	<0,0005							0,216	<0,0005	
1,00	0,805	0,195								0,195		
Continuous observation along the profile	0,365											

$$b'=0.3 \quad -30° \leqslant \Theta \leqslant 30° \quad d=1.2$$

h in units of d	P_0	P_1	P_2	P_3	P_4	P_5	P_6	P_7	$P_{\geqslant 8}$	$P_{\geqslant 1}$	$P_{\geqslant 2}$	Probability of interception by two profiles
0,05	0,206	0,030	0,065	0,111	0,202	0,344	0,041			0,794	0,764	
0,10	0,221	0,136	0,430	0,214						0,779	0,644	
0,15	0,248	0,414	0,338							0,752	0,338	
0,20	0,289	0,604	0,107							0,711	0,107	
0,25	0,354	0,638	0,008							0,646	0,008	
0,30	0,455	0,545								0,545		
0,35	0,533	0,467								0,467		
0,40	0,591	0,409								0,409		
0,45	0,636	0,364								0,364		
0,50	0,673	0,327								0,327		
0,60	0,727	0,273								0,273		
0,70	0,766	0,234								0,234		
0,80	0,795	0,205								0,205		
0,90	0,818	0,182								0,182		
1,00	0,836	0,164								0,164		
Continuous observation along the profile	0,201											

$$b'=0.3 \quad -90° \leqslant \Theta \leqslant 90° \quad d=1.2$$

h in units of d	P_0	P_1	P_2	P_3	P_4	P_5	P_6	P_7	$P_{\geqslant 8}$	$P_{\geqslant 1}$	$P_{\geqslant 2}$	Probability of interception by two profiles
0,05	0,421	0,016	0,033	0,055	0,095	0,160	0,077	0,040	0,103	0,579	0,564	
0,10	0,429	0,068	0,202	0,178	0,055	0,028	0,018	0,013	0,009	0,571	0,503	
0,15	0,442	0,199	0,251	0,062	0,028	0,016	0,002			0,558	0,359	
0,20	0,463	0,325	0,158	0,039	0,015	<0,0005				0,537	0,212	
0,25	0,494	0,385	0,094	0,025	0,001					0,506	0,121	
0,30	0,542	0,381	0,067	0,010						0,458	0,077	
0,35	0,587	0,362	0,049	0,003						0,413	0,052	
0,40	0,625	0,340	0,034	<0,0005						0,375	0,034	
0,45	0,659	0,319	0,022							0,341	0,022	
0,50	0,687	0,299	0,014							0,313	0,014	
0,60	0,732	0,262	0,005							0,268	0,005	
0,70	0,768	0,231	0,001							0,232	0,001	
0,80	0,796	0,204	<0,0005							0,205	<0,0005	
0,90	0,818	0,182								0,182		
1,00	0,836	0,164								0,164		
Continuous observation along the profile	0,418											

Table 87

b′=0,3   −30° ⩽ Θ ⩽ 30°   d=1,3

h in units of d	P_0	P_1	P_2	P_3	P_4	P_5	P_6	P_7	$P_{\geqslant 8}$	$P_{\geqslant 1}$	$P_{\geqslant 2}$	Probability of interception by two profiles
0,05	0,268	0,033	0,072	0,127	0,274	0,223	0,003			0,733	0,699	
0,10	0,284	0,152	0,450	0,114						0,716	0,564	
0,15	0,314	0,444	0,243							0,687	0,243	
0,20	0,360	0,583	0,057							0,640	0,057	
0,25	0,443	0,557	0,001							0,557	0,001	
0,30	0,535	0,465								0,465		
0,35	0,602	0,398								0,398		
0,40	0,651	0,349								0,349		
0,45	0,690	0,310								0,310		
0,50	0,721	0,279								0,279		
0,60	0,768	0,232								0,232		
0,70	0,801	0,199								0,199		
0,80	0,826	0,174								0,174		
0,90	0,845	0,155								0,155		
1,00	0,861	0,139								0,139		
Continuous observation along the profile	0,262											

Table 88

b′=0,3   −90° ⩽ Θ ⩽ 90°   d=1,3

h in units of d	P_0	P_1	P_2	P_3	P_4	P_5	P_6	P_7	$P_{\geqslant 8}$	$P_{\geqslant 1}$	$P_{\geqslant 2}$	Probability of interception by two profiles
0,05	0,466	0,017	0,036	0,063	0,123	0,131	0,054	0,031	0,079	0,534	0,517	
0,10	0,474	0,076	0,220	0,135	0,043	0,023	0,015	0,011	0,002	0,526	0,450	
0,15	0,489	0,218	0,210	0,049	0,023	0,010	<0,0005			0,511	0,293	
0,20	0,512	0,326	0,122	0,032	0,008					0,488	0,162	
0,25	0,551	0,358	0,074	0,017	<0,0005					0,449	0,091	
0,30	0,598	0,344	0,053	0,005						0,402	0,058	
0,35	0,640	0,322	0,037	0,001						0,360	0,038	
0,40	0,675	0,301	0,024							0,325	0,024	
0,45	0,705	0,281	0,015							0,295	0,015	
0,50	0,730	0,261	0,009							0,270	0,009	
0,60	0,770	0,227	0,003							0,230	0,003	
0,70	0,801	0,199	<0,0005							0,199	<0,0005	
0,80	0,826	0,174								0,174		
0,90	0,845	0,155								0,155		
1,00	0,861	0,139								0,139		
Continuous observation along the profile	0,463											

Table 89

$$b' = 0.3 \qquad -30° \leqslant \Theta \leqslant 30° \qquad d = 1,4$$

h in units of d	P_0	P_1	P_2	P_3	P_4	P_5	P_6	P_7	$P_{\geqslant 8}$	$P_{\geqslant 1}$	$P_{\geqslant 2}$	Probability of interception by two profiles
0,05	0,321	0,036	0,079	0,148	0,317	0,100				0,679	0,643	
0,10	0,339	0,171	0,441	0,050						0,661	0,491	
0,15	0,371	0,457	0,172							0,629	0,172	
0,20	0,424	0,551	0,025							0,576	0,025	
0,25	0,519	0,481								0,481		
0,30	0,599	0,401								0,401		
0,35	0,657	0,343								0,343		
0,40	0,699	0,301								0,301		
0,45	0,733	0,267								0,267		
0,50	0,760	0,240								0,240		
0,60	0,800	0,200								0,200		
0,70	0,828	0,172								0,172		
0,80	0,850	0,150								0,150		
0,90	0,866	0,134								0,134		
1,00	0,880	0,120								0,120		
Continuous observation along the profile	0,315											

Table 90

$$b' = 0,3 \qquad -90° \leqslant \Theta \leqslant 90° \qquad d = 1,4$$

h in units of d	P_0	P_1	P_2	P_3	P_4	P_5	P_6	P_7	$P_{\geqslant 8}$	$P_{\geqslant 1}$	$P_{\geqslant 2}$	Probability of interception by two profiles
0,05	0,504	0,018	0,040	0,072	0,143	0,095	0,041	0,025	0,062	0,496	0,477	
0,10	0,514	0,085	0,227	0,101	0,035	0,019	0,013	0,007	<0,0005	0,486	0,402	
0,15	0,530	0,232	0,174	0,040	0,020	0,005				0,470	0,239	
0,20	0,556	0,320	0,095	0,026	0,003					0,444	0,124	
0,25	0,600	0,329	0,060	0,011						0,400	0,071	
0,30	0,646	0,310	0,042	0,002						0,354	0,044	
0,35	0,684	0,289	0,027	<0,0005						0,316	0,027	
0,40	0,716	0,268	0,016							0,284	0,016	
0,45	0,742	0,248	0,010							0,258	0,010	
0,50	0,765	0,230	0,005							0,235	0,005	
0,60	0,801	0,198	0,001							0,199	0,001	
0,70	0,828	0,172	<0,0005							0,172	<0,0005	
0,80	0,850	0,150								0,150		
0,90	0,866	0,134								0,134		
1,00	0,880	0,120								0,120		
Continuous observation along the profile	0,501											

Table 91

$$b'=0,3 \qquad -30° \leqslant \Theta \leqslant 30° \qquad d=1,5$$

h in units of d	P_0	P_1	P_2	P_3	P_4	P_5	P_6	P_7	$P_{\geqslant 8}$	$P_{\geqslant 1}$	$P_{\geqslant 2}$	Probability of interception by two profiles
0,05	0,367	0,039	0,087	0,176	0,306	0,027				0,633	0,595	
0,10	0,386	0,194	0,407	0,013						0,614	0,420	
0,15	0,421	0,459	0,120							0,579	0,120	
0,20	0,483	0,510	0,007							0,517	0,007	
0,25	0,581	0,419								0,419		
0,30	0,651	0,349								0,349		
0,35	0,701	0,299								0,299		
0,40	0,738	0,262								0,262		
0,45	0,767	0,233								0,233		
0,50	0,791	0,209								0,209		
0,60	0,825	0,175								0,175		
0,70	0,850	0,150								0,150		
0,80	0,869	0,131								0,131		
0,90	0,884	0,116								0,116		
1,00	0,895	0,105								0,105		
Continuous observation along the profile	0,361											

Table 92

$$b'=0,3 \qquad -90° \leqslant \Theta \leqslant 90° \qquad d=1,5$$

h in units of d	P_0	P_1	P_2	P_3	P_4	P_5	P_6	P_7	$P_{\geqslant 8}$	$P_{\geqslant 1}$	$P_{\geqslant 2}$	Probability of interception by two profiles
0,05	0,538	0,020	0,044	0,084	0,148	0,066	0,032	0,020	0,049	0,462	0,442	
0,10	0,548	0,095	0,223	0,076	0,029	0,017	0,011	0,002		0,452	0,357	
0,15	0,566	0,240	0,144	0,034	0,015	0,001				0,435	0,195	
0,20	0,595	0,308	0,075	0,020	0,001					0,405	0,096	
0,25	0,643	0,302	0,049	0,006						0,357	0,056	
0,30	0,685	0,281	0,033	0,001						0,315	0,034	
0,35	0,721	0,260	0,020							0,279	0,020	
0,40	0,749	0,239	0,011							0,251	0,011	
0,45	0,773	0,220	0,006							0,227	0,006	
0,50	0,794	0,203	0,003							0,206	0,003	
0,60	0,826	0,174	<0,0005							0,174	<0,0005	
0,70	0,850	0,150								0,150		
0,80	0,869	0,131								0,131		
0,90	0,884	0,116								0,116		
1,00	0,895	0,105								0,105		
Continuous observation along the profile	0,535											

Table 93

b' = 0,3   —30° ⩽ Θ ⩽ 30°   d = 1,7

h in units of d	P_0	P_1	P_2	P_3	P_4	P_5	P_6	P_7	$P_{\geqslant 8}$	$P_{>1}$	$P_{>2}$	Probability of interception by two profiles
0,05	0,443	0,044	0,104	0,256	0,152	<0,0005				0,557	0,513	
0,10	0,465	0,255	0,280	<0,0005						0,535	0,280	
0,15	0,507	0,442	0,051							0,493	0,051	
0,20	0,592	0,408	<0,0005							0,408	<0,0005	
0,25	0,674	0,326								0,326		
0,30	0,728	0,272								0,272		
0,35	0,767	0,233								0,233		
0,40	0,796	0,204								0,204		
0,45	0,819	0,181								0,181		
0,50	0,837	0,163								0,163		
0,60	0,864	0,136								0,136		
0,70	0,884	0,116								0,116		
0,80	0,898	0,102								0,102		
0,90	0,909	0,091								0,091		
1,00	0,918	0,082								0,082		
Continuous observation along the profile	0,436											

Table 94

b' = 0,3   —90° ⩽ Θ ⩽ 90°   d = 1,7

h in units of d	P_0	P_1	P_2	P_3	P_4	P_5	P_6	P_7	$P_{\geqslant 8}$	$P_{>1}$	$P_{\geqslant 2}$	Probability of interception by two profiles
0,05	0,593	0,023	0,052	0,117	0,109	0,040	0,022	0,014	0,031	0,407	0,384	
0,10	0,604	0,121	0,187	0,049	0,021	0,013	0,004			0,396	0,274	
0,15	0,625	0,245	0,098	0,025	0,007					0,375	0,130	
0,20	0,665	0,272	0,052	0,010						0,335	0,062	
0,25	0,711	0,254	0,034	0,002						0,289	0,035	
0,30	0,748	0,233	0,020							0,252	0,020	
0,35	0,777	0,212	0,010							0,223	0,010	
0,40	0,801	0,193	0,005							0,199	0,005	
0,45	0,821	0,177	0,002							0,179	0,002	
0,50	0,838	0,161	0,001							0,162	0,001	
0,60	0,864	0,136								0,136		
0,70	0,884	0,116								0,116		
0,80	0,898	0,102								0,102		
0,90	0,909	0,091								0,091		
1,00	0,918	0,082								0,082		
Continuous observation along the profile	0,589											

59

Table 95

$b'=0,3$ $-30° \leqslant \Theta \leqslant 30°$ $d=2,0$

h in units of d	P_0	P_1	P_2	P_3	P_4	P_5	P_6	P_7	$P_{\geqslant 8}$	$P_{\geqslant 1}$	$P_{\geqslant 2}$	Probability of interception by two profiles
0,05	0,529	0,054	0,143	0,260	0,015					0,471	0,417	
0,10	0,556	0,300	0,145							0,444	0,145	
0,15	0,612	0,383	0,005							0,388	0,005	
0,20	0,705	0,295								0,295		
0,25	0,764	0,236								0,236		
0,30	0,804	0,196								0,196		
0,35	0,832	0,168								0,168		
0,40	0,853	0,147								0,147		
0,45	0,869	0,131								0,131		
0,50	0,882	0,118								0,118		
0,60	0,902	0,098								0,098		
0,70	0,916	0,084								0,084		
0,80	0,926	0,074								0,074		
0,90	0,935	0,065								0,065		
1,00	0,941	0,059								0,059		
Continuous observation along the profile	0,520											

Table 96

$b'=0,3$ $-90° \leqslant \Theta \leqslant 90°$ $d=2,0$

h in units of d	P_0	P_1	P_2	P_3	P_4	P_5	P_6	P_7	$P_{\geqslant 8}$	$P_{\geqslant 1}$	$P_{\geqslant 2}$	Probability of interception by two profiles
0,05	0,655	0,027	0,069	0,133	0,054	0,024	0,014	0,010	0,015	0,345	0,317	
0,10	0,669	0,149	0,132	0,031	0,015	0,005	<0,0005			0,331	0,182	
0,15	0,697	0,231	0,056	0,015	0,001					0,304	0,072	
0,20	0,743	0,221	0,033	0,003						0,257	0,035	
0,25	0,782	0,200	0,018							0,218	0,018	
0,30	0,812	0,179	0,008							0,188	0,008	
0,35	0,835	0,161	0,004							0,165	0,004	
0,40	0,854	0,145	0,001							0,146	0,001	
0,45	0,869	0,130	<0,0005							0,131	<0,0005	
0,50	0,882	0,118								0,118		
0,60	0,902	0,098								0,098		
0,70	0,916	0,084								0,084		
0,80	0,926	0,074								0,074		
0,90	0,935	0,065								0,065		
1,00	0,941	0,059								0,059		
Continuous observation along the profile	0,651											

Table 97

b′=0,2   —30° ≤ Θ ≤ 30°   d=0,5

h in units of d	P_0	P_1	P_2	P_3	P_4	P_5	P_6	P_7	$P_{>8}$	$P_{\geqslant 1}$	$P_{\geqslant 2}$	Probability of interception by two profiles
0,05								<0,005	1,00	1,00	1,00	0,91
0,10				<0,005	0,07	0,13	0,31	0,40	0,08	1,00	1,00	0,89
0,15			0,02	0,18	0,42	0,35	0,03			1,00	1,00	0,87
0,20		<0,005	0,16	0,54	0,29	<0,005				1,00	1,00	0,84
0,25		0,06	0,43	0,44	0,07					1,00	0,94	0,79
0,30		0,17	0,58	0,25	<0,005					1,00	0,83	0,73
0,35	<0,005	0,28	0,64	0,08						1,00	0,72	0,66
0,40	<0,005	0,42	0,57	<0,005						1,00	0,57	0,55
0,45	0,05	0,51	0,45							0,95	0,45	0,44
0,50	0,10	0,54	0,36							0,90	0,36	0,36
0,60	0,19	0,57	0,24							0,81	0,24	0,24
0,70	0,26	0,57	0,16							0,74	0,16	0,16
0,80	0,34	0,54	0,12							0,66	0,12	0,12
0,90	0,40	0,49	0,10							0,60	0,10	0,10
1,00	0,46	0,45	0,09							0,54	0,09	0,09
Continuous observation along the profile	—											0,91

Table 98

b′=0,2   —90° ≤ Θ ≤ 90°   d=0,5

h in units of d	P_0	P_1	P_2	P_3	P_4	P_5	P_6	P_7	$P_{>8}$	$P_{\geqslant 1}$	$P_{\geqslant 2}$	Probability of interception by two profiles
0,05	0,11	<0,005	<0,005	<0,005	<0,005	<0,005	<0,005	<0,005	0,88	0,89	0,89	0,45
0,10	0,11	<0,005	<0,005	0,01	0,04	0,15	0,21	0,25	0,22	0,89	0,89	0,44
0,15	0,11	<0,005	0,02	0,15	0,31	0,26	0,07	0,02	0,06	0,89	0,88	0,43
0,20	0,11	0,01	0,13	0,40	0,24	0,05	0,02	0,02	0,03	0,89	0,88	0,41
0,25	0,11	0,04	0,34	0,36	0,09	0,03	0,02	0,01	0,01	0,89	0,85	0,39
0,30	0,11	0,12	0,47	0,22	0,04	0,02	0,01	<0,005		0,89	0,77	0,36
0,35	0,11	0,22	0,50	0,12	0,03	0,02	<0,005			0,89	0,67	0,33
0,40	0,12	0,33	0,45	0,06	0,03	0,01				0,88	0,55	0,28
0,45	0,13	0,42	0,38	0,04	0,02	<0,005				0,87	0,44	0,23
0,50	0,16	0,48	0,31	0,04	0,01					0,84	0,35	0,19
0,60	0,23	0,53	0,22	0,03	<0,005					0,77	0,24	0,13
0,70	0,29	0,53	0,16	0,01						0,71	0,18	0,10
0,80	0,35	0,52	0,13	<0,005						0,65	0,13	0,07
0,90	0,40	0,49	0,10	<0,005						0,60	0,10	0,06
1,00	0,45	0,47	0,08							0,55	0,08	0,05
Continuous observation along the profile	0,11											0,45

Table 99

b′ = 0,2   —30° ⩽ Θ ⩽ 30°   d = 0,6

h in units of d	P_0	P_1	P_2	P_3	P_4	P_5	P_6	P_7	$P_{\geqslant 8}$	$P_{\geqslant 1}$	$P_{\geqslant 2}$	Probability of interception by two profiles
0,05							0,06	0,32	0,61	1,00	1,00	0,59
0,10				0,23	0,31	0,33	0,13	<0,005		1,00	1,00	0,57
0,15			0,32	0,45	0,22	0,01				1,00	1,00	0,54
0,20		0,14	0,56	0,27	0,02					1,00	0,86	0,50
0,25		0,35	0,56	0,09						1,00	0,65	0,44
0,30		0,56	0,43	0,01						1,00	0,44	0,37
0,35	0,04	0,68	0,28							0,96	0,28	0,28
0,40	0,12	0,66	0,22							0,88	0,22	0,22
0,45	0,20	0,62	0,17							0,80	0,17	0,17
0,50	0,27	0,60	0,14							0,73	0,14	0,14
0,60	0,36	0,55	0,09							0,64	0,09	0,09
0,70	0,43	0,51	0,06							0,57	0,06	0,06
0,80	0,50	0,45	0,05							0,50	0,05	0,05
0,90	0,56	0,40	0,04							0,44	0,04	0,04
1,00	0,60	0,36	0,04							0,40	0,04	0,04
Continuous observation along the profile	—											0,59

Table 100

b′ = 0,2   —90° ⩽ Θ ⩽ 90°   d = 0,6

h in units of d	P_0	P_1	P_2	P_3	P_4	P_5	P_6	P_7	$P_{\geqslant 8}$	$P_{\geqslant 1}$	$P_{\geqslant 2}$	Probability of interception by two profiles
0,05	0,15	<0,005	<0,005	<0,005	0,01	0,01	0,03	0,15	0,65	0,85	0,85	0,26
0,10	0,15	<0,005	0,01	0,12	0,26	0,23	0,11	0,03	0,08	0,85	0,84	0,25
0,15	0,15	0,01	0,19	0,36	0,18	0,04	0,02	0,02	0,03	0,85	0,84	0,24
0,20	0,15	0,08	0,43	0,23	0,05	0,02	0,02	0,01	0,01	0,85	0,77	0,22
0,25	0,16	0,23	0,44	0,11	0,03	0,02	0,01			0,84	0,62	0,20
0,30	0,16	0,38	0,37	0,06	0,02	0,01				0,84	0,46	0,16
0,35	0,18	0,49	0,27	0,04	0,02					0,82	0,33	0,12
0,40	0,22	0,53	0,20	0,03	0,01					0,78	0,25	0,09
0,45	0,26	0,54	0,16	0,03	0,01					0,74	0,19	0,07
0,50	0,31	0,54	0,13	0,02						0,69	0,16	0,06
0,60	0,39	0,51	0,10	0,01						0,61	0,11	0,04
0,70	0,45	0,47	0,07	<0,005						0,55	0,07	0,03
0,80	0,51	0,44	0,05	<0,005						0,49	0,05	0,02
0,90	0,55	0,41	0,04							0,45	0,04	0,02
1,00	0,59	0,38	0,03							0,41	0,03	0,01
Continuous observation along the profile	0,15											0,27

Table 101

$$b'=0,2 \qquad -30° \leqslant \Theta \leqslant 30° \qquad d=0,7$$

h in units of d	P_0	P_1	P_2	P_3	P_4	P_5	P_6	P_7	$P_{\geqslant 8}$	$P_{\geqslant 1}$	$P_{\geqslant 2}$	Probability of interception by two profiles
0,05						0,18	0,47	0,15	0,20	1,00	1,00	0,36
0,10			0,10	0,64	0,22	0,04	<0,005			1,00	1,00	0,34
0,15		0,08	0,72	0,19	0,01					1,00	0,92	0,31
0,20		0,44	0,52	0,04	<0,005					1,00	0,56	0,26
0,25	0,01	0,70	0,29	<0,005						0,99	0,29	0,19
0,30	0,07	0,80	0,13							0,93	0,13	0,13
0,35	0,18	0,73	0,10							0,82	0,10	0,10
0,40	0,28	0,65	0,08							0,72	0,08	0,08
0,45	0,35	0,59	0,06							0,65	0,06	0,06
0,50	0,41	0,54	0,05							0,59	0,05	0,05
0,60	0,49	0,48	0,03							0,51	0,03	0,03
0,70	0,56	0,42	0,02							0,44	0,02	0,02
0,80	0,62	0,37	0,02							0,38	0,02	0,02
0,90	0,66	0,33	0,02							0,34	0,02	0,02
1,00	0,69	0,29	0,01							0,31	0,01	0,01
Continuous observation along the profile	—											0,37

Table 102

$$b'=0,2 \qquad -90° \leqslant \Theta \leqslant 90° \qquad d=0,7$$

h in units of d	P_0	P_1	P_2	P_3	P_4	P_5	P_6	P_7	$P_{\geqslant 8}$	$P_{\geqslant 1}$	$P_{\geqslant 2}$	Probability of interception by two profiles
0,05	0,19	<0,005	0,01	0,01	0,02	0,10	0,23	0,15	0,30	0,81	0,81	0,14
0,10	0,19	0,01	0,07	0,36	0,20	0,07	0,03	0,02	0,05	0,81	0,80	0,14
0,15	0,19	0,06	0,43	0,19	0,05	0,02	0,02	0,01	0,02	0,81	0,74	0,12
0,20	0,20	0,26	0,40	0,08	0,03	0,02	0,01	0,01	<0,005	0,80	0,54	0,10
0,25	0,21	0,44	0,27	0,04	0,02	0,01	<0,005			0,79	0,35	0,07
0,30	0,24	0,53	0,17	0,03	0,02	<0,005				0,76	0,23	0,05
0,35	0,29	0,54	0,13	0,03	0,01	<0,005				0,71	0,16	0,04
0,40	0,35	0,53	0,10	0,02	<0,005					0,65	0,12	0,03
0,45	0,40	0,50	0,08	0,02	<0,005					0,60	0,10	0,02
0,50	0,44	0,48	0,07	0,01						0,56	0,08	0,02
0,60	0,52	0,43	0,05	<0,005						0,48	0,05	0,01
0,70	0,57	0,39	0,03	<0,005						0,43	0,03	0,01
0,80	0,62	0,36	0,02							0,38	0,02	0,01
0,90	0,66	0,33	0,01							0,34	0,01	0,01
1,00	0,69	0,30	0,01							0,31	0,01	<0,005
Continuous observation along the profile	0,19											0,15

Table 103

$$b'=0,2 \qquad -30° \leqslant \Theta \leqslant 30° \qquad d=0,8$$

h in units of d	P_0	P_1	P_2	P_3	P_4	P_5	P_6	P_7	$P_{\geqslant 8}$	$P_{\geqslant 1}$	$P_{\geqslant 2}$	Probability of interception by two profiles
0,05				0,01	0,24	0,60	0,15	0,01	—	1,00	1,00	0,19
0,10		0,01	0,54	0,43	0,01					1,00	0,99	0,17
0,15	<0,005	0,38	0,61	0,01						1,00	0,62	0,13
0,20	0,01	0,75	0,24							0,99	0,24	0,08
0,25	0,08	0,85	0,06							0,92	0,06	0,05
0,30	0,22	0,75	0,03							0,78	0,03	0,03
0,35	0,32	0,65	0,03							0,68	0,03	0,03
0,40	0,41	0,57	0,02							0,59	0,02	0,02
0,45	0,47	0,51	0,02							0,53	0,02	0,02
0,50	0,52	0,47	0,01							0,48	0,01	0,01
0,60	0,60	0,39	0,01							0,40	0,01	0,01
0,70	0,66	0,34	0,01							0,34	0,01	0,01
0,80	0,70	0,30	0,01							0,30	0,01	0,01
0,90	0,73	0,26	<0,005							0,27	<0,005	<0,005
1,00	0,76	0,24	<0,005							0,24	<0,005	<0,005
Continuous observation along the profile	—											0,20

Table 104

$$b'=0,2 \qquad -90° \leqslant \Theta \leqslant 90° \qquad d=0,8$$

h in units of d	P_0	P_1	P_2	P_3	P_4	P_5	P_6	P_7	$P_{\geqslant 8}$	$P_{\geqslant 1}$	$P_{\geqslant 2}$	Probability of interception by two profiles
0,05	0,23	0,01	0,01	0,03	0,12	0,26	0,14	0,06	0,15	0,77	0,76	0,07
0,10	0,24	0,03	0,26	0,29	0,08	0,04	0,02	0,01	0,03	0,76	0,73	0,06
0,15	0,24	0,20	0,39	0,09	0,03	0,02	0,01	0,01	0,01	0,76	0,55	0,04
0,20	0,26	0,41	0,25	0,05	0,02	0,01	0,01	<0,005	—	0,74	0,33	0,03
0,25	0,29	0,52	0,14	0,03	0,02	0,01	<0,005			0,71	0,19	0,02
0,30	0,35	0,52	0,09	0,02	0,01	<0,005				0,65	0,13	0,01
0,35	0,42	0,49	0,07	0,02	<0,005					0,58	0,09	0,01
0,40	0,47	0,46	0,06	0,01	<0,005					0,53	0,07	0,01
0,45	0,51	0,43	0,05	0,01						0,49	0,05	0,01
0,50	0,55	0,41	0,04	<0,005						0,45	0,04	<0,005
0,60	0,62	0,36	0,02	<0,005						0,38	0,02	<0,005
0,70	0,66	0,32	0,01							0,34	0,01	<0,005
0,80	0,70	0,29	0,01							0,30	0,01	<0,005
0,90	0,73	0,26	<0,005							0,27	<0,005	<0,005
1,00	0,76	0,24	<0,005							0,24	<0,005	<0,005
Continuous observation along the profile	0,23											0,07

64

Table 105

$$b'=0,2 \qquad -30° \leqslant \Theta \leqslant 30° \qquad d=0,9$$

h in units of d	P_0	P_1	P_2	P_3	P_4	P_5	P_6	P_7	$P_{\geqslant 8}$	$P_{\geqslant 1}$	$P_{\geqslant 2}$	Probability of interception by two profiles
0,05	<0,005	0,01	0,05	0,22	0,47	0,24	<0,005			1,00	0,99	0,06
0,10	0,01	0,17	0,70	0,12						0,99	0,82	0,04
0,15	0,03	0,64	0,33							0,97	0,33	0,02
0,20	0,10	0,83	0,07							0,90	0,07	0,01
0,25	0,23	0,76	0,01							0,77	0,01	0,01
0,30	0,36	0,64	<0,005							0,64	<0,005	<0,005
0,35	0,45	0,55	<0,005							0,55	<0,005	<0,005
0,40	0,52	0,48	<0,005							0,48	<0,005	<0,005
0,45	0,57	0,43	<0,005							0,43	<0,005	<0,005
0,50	0,61	0,38	<0,005							0,39	<0,005	<0,005
0,60	0,68	0,32	<0,005							0,32	<0,005	<0,005
0,70	0,72	0,27	<0,005							0,28	<0,005	<0,005
0,80	0,76	0,24	<0,005							0,24	<0,005	<0,005
0,90	0,79	0,21	<0,005							0,21	<0,005	<0,005
1,00	0,81	0,19	<0,005							0,19	<0,005	<0,005
Continuous observation along the profile	<0,005											0,06

Table 106

$$b'=0,2 \qquad -90° \leqslant \Theta \leqslant 90° \qquad d=0,9$$

h in units of d	P_0	P_1	P_2	P_3	P_4	P_5	P_6	P_7	$P_{\geqslant 8}$	$P_{\geqslant 1}$	$P_{\geqslant 2}$	Probability of interception by two profiles
0,05	0,28	0,01	0,04	0,10	0,20	0,16	0,07	0,04	0,10	0,72	0,71	0,02
0,10	0,29	0,09	0,33	0,17	0,05	0,03	0,01	0,01	0,02	0,71	0,62	0,01
0,15	0,30	0,31	0,28	0,06	0,02	0,01	0,01	0,01	<0,005	0,70	0,38	0,01
0,20	0,34	0,46	0,15	0,03	0,02	0,01	<0,005			0,67	0,21	<0,005
0,25	0,39	0,49	0,09	0,02	0,01	<0,005				0,61	0,12	<0,005
0,30	0,46	0,46	0,06	0,02	<0,005					0,54	0,08	<0,005
0,35	0,52	0,42	0,05	0,01	<0,005					0,48	0,06	<0,005
0,40	0,56	0,39	0,04	0,01						0,44	0,04	<0,005
0,45	0,60	0,36	0,03	<0,005						0,40	0,03	<0,005
0,50	0,64	0,34	0,02	<0,005						0,36	0,02	<0,005
0,60	0,69	0,30	0,01							0,31	0,01	<0,005
0,70	0,73	0,26	0,01							0,27	0,01	<0,005
0,80	0,76	0,24	<0,005							0,24	<0,005	<0,005
0,90	0,79	0,21	<0,005							0,21	<0,005	<0,005
1,00	0,81	0,19	<0,005							0,19	<0,005	<0,005
Continuous observation along the profile	0,28											0,02

Table 107

$$b'=0,2 \qquad -30° \leqslant \Theta \leqslant 30° \qquad d=1,0$$

h in units of d	P_0	P_1	P_2	P_3	P_4	P_5	P_6	P_7	$P_{\geqslant 8}$	$P_{\geqslant 1}$	$P_{\geqslant 2}$	Probability of interception by two profiles
0,05	0,053	0,058	0,129	0,261	0,456	0,043				0,947	0,890	
0,10	0,081	0,288	0,609	0,022						0,919	0,630	
0,15	0,134	0,685	0,181							0,866	0,181	
0,20	0,226	0,764	0,011							0,774	0,011	
0,25	0,372	0,628								0,628		
0,30	0,477	0,523								0,523		
0,35	0,551	0,449								0,449		
0,40	0,607	0,393								0,393		
0,45	0,651	0,349								0,349		
0,50	0,686	0,314								0,314		
0,60	0,738	0,262								0,262		
0,70	0,776	0,224								0,224		
0,80	0,804	0,196								0,196		
0,90	0,826	0,174								0,174		
1,00	0,843	0,157								0,157		
Continuous observation along the profile	0,043											

Table 108

$$b'=0,2 \qquad -90° \leqslant \Theta \leqslant 90° \qquad d=1,0$$

h in units of d	P_0	P_1	P_2	P_3	P_4	P_5	P_6	P_7	$P_{\geqslant 8}$	$P_{\geqslant 1}$	$P_{\geqslant 2}$	Probability of interception by two profiles
0,05	0,336	0,028	0,063	0,121	0,213	0,094	0,044	0,026	0,074	0,663	0,635	
0,10	0,350	0,137	0,321	0,104	0,036	0,019	0,012	0,008	0,013	0,650	0,513	
0,15	0,376	0,345	0,201	0,041	0,018	0,011	0,007	0,001		0,624	0,279	
0,20	0,419	0,441	0,098	0,025	0,013	0,005				0,581	0,140	
0,25	0,487	0,428	0,060	0,019	0,006					0,513	0,085	
0,30	0,548	0,394	0,044	0,013	0,001					0,452	0,057	
0,35	0,598	0,361	0,034	0,006						0,401	0,040	
0,40	0,639	0,331	0,027	0,002						0,361	0,030	
0,45	0,673	0,306	0,021	<0,0005						0,327	0,021	
0,50	0,701	0,284	0,015							0,299	0,015	
0,60	0,745	0,247	0,007							0,255	0,007	
0,70	0,779	0,218	0,003							0,221	0,003	
0,80	0,805	0,194	0,001							0,195	0,001	
0,90	0,826	0,174	<0,0005							0,174	<0,0005	
1,00	0,843	0,157								0,157		
Continuous observation along the profile	0,331											

Table 109

$$b'=0,2 \qquad -30° \leqslant \Theta \leqslant 30° \qquad d=1,1$$

h in units of d	P_0	P_1	P_2	P_3	P_4	P_5	P_6	P_7	$P_{\geqslant 8}$	$P_{\geqslant 1}$	$P_{\geqslant 2}$	Probability of interception by two profiles
0,05	0,140	0,064	0,148	0,355	0,290	0,002				0,860	0,796	
0,10	0,172	0,358	0,469	0,001						0,828	0,470	
0,15	0,232	0,670	0,098							0,768	0,098	
0,20	0,351	0,648	<0,0005							0,649	<0,0005	
0,25	0,481	0,519								0,519		
0,30	0,567	0,433								0,433		
0,35	0,629	0,371								0,371		
0,40	0,675	0,325								0,325		
0,45	0,712	0,288								0,288		
0,50	0,740	0,260								0,260		
0,60	0,784	0,216								0,216		
0,70	0,815	0,185								0,185		
0,80	0,838	0,162								1,162		
0,90	0,856	0,144								0,144		
1,00	0,870	0,130								0,130		
Continuous observation along the profile	0,130											

Table 110

$$b'=0,2 \qquad -90° \leqslant \Theta \leqslant 90° \qquad d=1,1$$

h in units of d	P_0	P_1	P_2	P_3	P_4	P_5	P_6	P_7	$P_{\geqslant 8}$	$P_{\geqslant 1}$	$P_{\geqslant 2}$	Probability of interception by two profiles
0,05	0,397	0,032	0,071	0,158	0,173	0,062	0,032	0,020	0,056	0,603	0,572	
0,10	0,413	0,166	0,283	0,073	0,028	0,015	0,009	0,007	0,007	0,587	0,421	
0,15	0,442	0,352	0,148	0,031	0,014	0,009	0,004	<0,0005		0,558	0,206	
0,20	0,496	0,402	0,071	0,020	0,010	0,001				0,504	0,102	
0,25	0,563	0,374	0,046	0,015	0,002					0,437	0,063	
0,30	0,618	0,340	0,034	0,008	<0,0005					0,382	0,042	
0,35	0,662	0,309	0,027	0,003						0,338	0,030	
0,40	0,697	0,282	0,020	0,001						0,303	0,021	
0,45	0,726	0,260	0,014	<0,0005						0,274	0,014	
0,50	0,750	0,240	0,010							0,250	0,010	
0,60	0,788	0,208	0,004							0,212	0,004	
0,70	0,816	0,183	0,001							0,184	0,001	
0,80	0,838	0,162	<0,0005							0,162	<0,0005	
0,90	0,856	0,144	<0,0005							0,144	<0,0005	
1,00	0,870	0,130								0,130		
Continuous observation along the profile	0,392											

Table 111

$$b'=0,2 \qquad -30° \leqslant \Theta \leqslant 30° \qquad d=1,2$$

h in units of d	P_0	P_1	P_2	P_3	P_4	P_5	P_6	P_7	$P_{\geqslant 8}$	$P_{\geqslant 1}$	$P_{\geqslant 2}$	Probability of interception by two profiles
0,05	0,214	0,071	0,171	0,409	0,135						0,786	0,715
0,10	0,249	0,411	0,340								0,751	0,340
0,15	0,318	0,637	0,045								0,682	0,045
0,20	0,455	0,545									0,545	
0,25	0,564	0,436									0,436	
0,30	0,636	0,364									0,364	
0,35	0,688	0,312									0,312	
0,40	0,727	0,273									0,273	
0,45	0,758	0,242									0,242	
0,50	0,782	0,218									0,218	
0,60	0,818	0,182									0,182	
0,70	0,844	0,156									0,156	
0,80	0,864	0,136									0,136	
0,90	0,879	0,121									0,121	
1,00	0,891	0,109									0,109	
Continuous observation along the profile	0,203											

Table 112

$$b'=0,2 \qquad -90° \leqslant \Theta \leqslant 90° \qquad d=1,2$$

h in units of d	P_0	P_1	P_2	P_3	P_4	P_5	P_6	P_7	$P_{\geqslant 8}$	$P_{\geqslant 1}$	$P_{\geqslant 2}$	Probability of interception by two profiles
0,05	0,448	0,035	0,082	0,184	0,124	0,045	0,024	0,015	0,043	0,552	0,517	
0,10	0,466	0,191	0,239	0,054	0,021	0,012	0,008	0,006	0,004	0,534	0,343	
0,15	0,499	0,348	0,109	0,025	0,012	0,007	0,001			0,501	0,153	
0,20	0,561	0,362	0,054	0,016	0,006	<0,0005				0,439	0,077	
0,25	0,624	0,329	0,036	0,011	0,001					0,376	0,048	
0,30	0,673	0,295	0,027	0,004						0,327	0,032	
0,35	0,711	0,267	0,021	0,001						0,289	0,022	
0,40	0,742	0,243	0,015	<0,0005						0,258	0,015	
0,45	0,767	0,223	0,009							0,233	0,009	
0,50	0,788	0,206	0,006							0,212	0,006	
0,60	0,820	0,177	0,002							0,180	0,002	
0,70	0,845	0,155	0,001							0,155	0,001	
0,80	0,864	0,136	<0,0005							0,136	<0,0005	
0,90	0,879	0,121								0,121		
1,00	0,891	0,109								0,109		
Continuous observation along the profile	0,443											

Table 113

b'=0,2   —30° ≤ Θ ≤ 30°   d=1,3

h in units of d	P_0	P_1	P_2	P_3	P_4	P_5	P_6	P_7	$P_{\geqslant 8}$	$P_{\geqslant 1}$	$P_{\geqslant 2}$	Probability of interception by two profiles
0,05	0,276	0,078	0,200	0,404	0,043					0,724	0,646	
0,10	0,315	0,441	0,244							0,685	0,244	
0,15	0,395	0,591	0,014							0,606	0,014	
0,20	0,535	0,465								0,465		
0,25	0,628	0,372								0,372		
0,30	0,690	0,310								0,310		
0,35	0,734	0,266								0,266		
0,40	0,768	0,232								0,232		
0,45	0,793	0,207								0,207		
0,50	0,814	0,186								0,186		
0,60	0,845	0,155								0,155		
0,70	0,867	0,133								0,133		
0,80	0,884	0,116								0,116		
0,90	0,897	0,103								0,103		
1,00	0,907	0,093								0,093		
Continuous observation along the profile	0,264											

Table 114

b'=0,2   —90° ≤ Θ ≤ 90°   d=1,3

h in units of d	P_0	P_1	P_2	P_3	P_4	P_5	P_6	P_7	$P_{\geqslant 8}$	$P_{\geqslant 1}$	$P_{\geqslant 2}$	Probability of interception by two profiles
0,05	0,492	0,038	0,094	0,193	0,085	0,034	0,019	0,012	0,033	0,509	0,470	
0,10	0,511	0,210	0,199	0,042	0,017	0,010	0,006	0,005	0,001	0,489	0,280	
0,15	0,548	0,337	0,081	0,020	0,010	0,004	<0,0005			0,452	0,115	
0,20	0,616	0,324	0,043	0,014	0,003					0,384	0,060	
0,25	0,673	0,290	0,030	0,007	<0,0005					0,327	0,037	
0,30	0,717	0,259	0,022	0,002						0,283	0,024	
0,35	0,751	0,233	0,016	<0,0005						0,249	0,016	
0,40	0,778	0,212	0,010							0,222	0,010	
0,45	0,800	0,194	0,006							0,200	0,006	
0,50	0,818	0,178	0,004							0,182	0,004	
0,60	0,846	0,153	0,001							0,154	0,001	
0,70	0,867	0,132	<0,0005							0,133	<0,0005	
0,80	0,884	0,116								0,116		
0,90	0,897	0,103								0,103		
1,00	0,907	0,093								0,093		
Continuous observation along the profile	0,486											

Table 115

$b'=0,2$ $-30° \leqslant \Theta \leqslant 30°$ $d=1,4$

h in units of d	P_0	P_1	P_2	P_3	P_4	P_5	P_6	P_7	$P_{\geqslant 8}$	$P_{\geqslant 1}$	$P_{\geqslant 2}$	Probability of interception by two profiles
0,05	0,330	0,085	0,245	0,334	0,007					0,670	0,585	
0,10	0,372	0,454	0,174							0,628	0,174	
0,15	0,468	0,530	0,002							0,532	0,002	
0,20	0,599	0,401								0,401		
0,25	0,679	0,321								0,321		
0,30	0,733	0,267								0,267		
0,35	0,771	0,229								0,229		
0,40	0,800	0,200								0,200		
0,45	0,822	0,178								0,178		
0,50	0,840	0,160								0,160		
0,60	0,866	0,134								0,134		
0,70	0,886	0,114								0,114		
0,80	0,900	0,100								0,100		
0,90	0,911	0,089								0,089		
1,00	0,920	0,080								0,080		
Continuous observation along the profile	0,317											

Table 116

$b'=0,2$ $-90° \leqslant \Theta \leqslant 90°$ $d=1,4$

h in units of d	P_0	P_1	P_2	P_3	P_4	P_5	P_6	P_7	$P_{\geqslant 8}$	$P_{\geqslant 1}$	$P_{\geqslant 2}$	Probability of interception by two profiles
0,05	0,529	0,042	0,112	0,178	0,061	0,027	0,015	0,010	0,026	0,471	0,430	
0,10	0,550	0,222	0,164	0,033	0,014	0,008	0,006	0,003	<0,0005	0,450	0,228	
0,15	0,594	0,317	0,062	0,017	0,008	0,002				0,406	0,089	
0,20	0,661	0,292	0,035	0,011	0,001					0,339	0,048	
0,25	0,713	0,258	0,025	0,005						0,287	0,029	
0,30	0,752	0,229	0,018	0,001						0,248	0,019	
0,35	0,783	0,206	0,012	<0,0005						0,217	0,012	
0,40	0,807	0,186	0,007							0,193	0,007	
0,45	0,826	0,170	0,004							0,174	0,004	
0,50	0,842	0,156	0,002							0,158	0,002	
0,60	0,867	0,133	<0,0005							0,133	<0,0005	
0,70	0,886	0,114	<0,0005							0,115	<0,0005	
0,80	0,900	0,100								0,100		
0,90	0,911	0,089								0,089		
1,00	0,920	0,080								0,080		
Continuous observation along the profile	0,522											

Table 117

b′=0,2   —30° ⩽ Θ ⩽ 30°   d=1,5

h in units of d	P_0	P_1	P_2	P_3	P_4	P_5	P_6	P_7	$P_{\geqslant 8}$	$P_{\geqslant 1}$	$P_{\geqslant 2}$	Probability of interception by two profiles
0,05	0,376	0,093	0,290	0,241	<0,0005					0,624	0,531	
0,10	0,423	0,457	0,121							0,577	0,121	
0,15	0,535	0,465	<0,0005							0,465	<0,0005	
0,20	0,651	0,349								0,349		
0,25	0,721	0,279								0,279		
0,30	0,767	0,233								0,233		
0,35	0,801	0,199								0,199		
0,40	0,825	0,175								0,175		
0,45	0,845	0,155								0,155		
0,50	0,860	0,140								0,140		
0,60	0,884	0,116								0,116		
0,70	0,900	0,100								0,100		
0,80	0,913	0,087								0,087		
0,90	0,922	0,078								0,078		
1,00	0,930	0,070								0,070		
Continuous observation along the profile	0,362											

Table 118

b′=0,2   —90° ⩽ Θ ⩽ 90°   d=1,5

h in units of d	P_0	P_1	P_2	P_3	P_4	P_5	P_6	P_7	$P_{\geqslant 8}$	$P_{\geqslant 1}$	$P_{\geqslant 2}$	Probability of interception by two profiles
0,05	0,561	0,045	0,131	0,153	0,047	0,022	0,012	0,008	0,020	0,439	0,394	
0,10	0,584	0,230	0,134	0,027	0,012	0,007	0,005	0,001		0,416	0,186	
0,15	0,635	0,294	0,050	0,014	0,007	0,001				0,365	0,071	
0,20	0,699	0,263	0,029	0,009	<0,0005					0,301	0,038	
0,25	0,747	0,230	0,021	0,003						0,253	0,023	
0,30	0,782	0,204	0,014	<0,0005						0,218	0,014	
0,35	0,809	0,183	0,008							0,191	0,008	
0,40	0,830	0,165	0,005							0,170	0,005	
0,45	0,847	0,150	0,003							0,153	0,003	
0,50	0,862	0,137	0,001							0,138	0,001	
0,60	0,884	0,116	<0,0005							0,116	<0,0005	
0,70	0,900	0,100								0,100		
0,80	0,913	0,087								0,087		
0,90	0,922	0,078								0,078		
1,00	0,930	0,070								0,070		
Continuous observation along the profile	0,554											

Table 119

$b'=0,2$ $\quad$ $-30° \leqslant \Theta \leqslant 30°$ $\quad$ $d=1,7$

h in units of d	P_0	P_1	P_2	P_3	P_4	P_5	P_6	P_7	$P_{\geqslant 8}$	$P_{\geqslant 1}$	$P_{\geqslant 2}$	Probability of interception by two profiles
0,05	0,453	0,110	0,334	0,103						0,547	0,437	
0,10	0,508	0,440	0,052							0,492	0,052	
0,15	0,638	0,362								0,362		
0,20	0,728	0,272								0,272		
0,25	0,783	0,217								0,217		
0,30	0,819	0,181								0,181		
0,35	0,845	0,155								0,155		
0,40	0,864	0,136								0,136		
0,45	0,879	0,121								0,121		
0,50	0,891	0,109								0,109		
0,60	0,909	0,091								0,091		
0,70	0,922	0,078								0,078		
0,80	0,932	0,068								0,068		
0,90	0,940	0,060								0,060		
1,00	0,946	0,054								0,054		
Continuous observation along the profile	0,437											

Table 120

$b'=0,2$ $\quad$ $-90° \leqslant \Theta \leqslant 90°$ $\quad$ $d=1,7$

h in units of d	P_0	P_1	P_2	P_3	P_4	P_5	P_6	P_7	$P_{\geqslant 8}$	$P_{\geqslant 1}$	$P_{\geqslant 2}$	Probability of interception by two profiles
0,05	0,615	0,053	0,156	0,103	0,030	0,015	0,009	0,006	0,013	0,385	0,332	
0,10	0,641	0,234	0,089	0,019	0,009	0,005	0,002			0,359	0,125	
0,15	0,702	0,250	0,034	0,011	0,003					0,298	0,048	
0,20	0,758	0,216	0,021	0,004						0,242	0,026	
0,25	0,798	0,187	0,014	0,001						0,202	0,015	
0,30	0,827	0,165	0,008							0,173	0,008	
0,35	0,849	0,146	0,004							0,151	0,004	
0,40	0,866	0,131	0,002							0,134	0,002	
0,45	0,880	0,119	0,001							0,120	0,001	
0,50	0,892	0,108	<0,0005							0,108	<0,0005	
0,60	0,909	0,091								0,091		
0,70	0,922	0,078								0,078		
0,80	0,932	0,068								0,068		
0,90	0,940	0,060								0,060		
1,00	0,946	0,054								0,054		
Continuous observation along the profile	0,607											

Table 121

$$b'=0,2 \qquad -30° \leqslant \Theta \leqslant 30° \qquad d=2,0$$

h in units of d	P_0	P_1	P_2	P_3	P_4	P_5	P_6	P_7	$P_{\geqslant 8}$	$P_{\geqslant 1}$	$P_{\geqslant 2}$	Probability of interception by two profiles
0,05	0,541	0,144	0,304	0,011						0,459	0,315	
0,10	0,613	0,382	0,005							0,387	0,005	
0,15	0,738	0,262								0,262		
0,20	0,804	0,196								0,196		
0,25	0,843	0,157								0,157		
0,30	0,869	0,131								0,131		
0,35	0,888	0,112								0,112		
0,40	0,902	0,098								0,098		
0,45	0,913	0,087								0,087		
0,50	0,921	0,079								0,079		
0,60	0,935	0,065								0,065		
0,70	0,944	0,056								0,056		
0,80	0,951	0,049								0,049		
0,90	0,956	0,044								0,044		
1,00	0,961	0,039								0,039		
Continuous observation along the profile	0,522											

Table 122

$$b'=0,2 \qquad -90° \leqslant \Theta \leqslant 90° \qquad d=2,0$$

h in units of d	P_0	P_1	P_2	P_3	P_4	P_5	P_6	P_7	$P_{\geqslant 8}$	$P_{\geqslant 1}$	$P_{\geqslant 2}$	Probability of interception by two profiles
0,05	0,675	0,069	0,160	0,052	0,018	0,009	0,006	0,004	0,006	0,325	0,256	
0,10	0,709	0,221	0,049	0,013	0,006	0,002	<0,0005			0,291	0,070	
0,15	0,774	0,197	0,022	0,007	<0,0005					0,226	0,029	
0,20	0,820	0,166	0,014	0,001						0,180	0,015	
0,25	0,851	0,142	0,008							0,149	0,008	
0,30	0,873	0,124	0,004							0,127	0,004	
0,35	0,889	0,109	0,002							0,111	0,002	
0,40	0,902	0,097	0,001							0,098	0,001	
0,45	0,913	0,087	<0,0005							0,087	<0,0005	
0,50	0,921	0,079								0,079		
0,60	0,935	0,065								0,065		
0,70	0,944	0,056								0,056		
0,80	0,951	0,049								0,049		
0,90	0,956	0,044								0,044		
1,00	0,961	0,039								0,039		
Continuous observation along the profile	0,666											

Table 123

$$b'=0,1 \quad -30° \leqslant \Theta \leqslant 30° \quad d=0,5$$

h in units of d	P_0	P_1	P_2	P_3	P_4	P_5	P_6	P_7	$P_{\geqslant 8}$	$P_{\geqslant 1}$	$P_{\geqslant 2}$	Probability of interception by two profiles
0,05				<0,005	0,07	0,14	0,31	0,39	0,09	1,00	1,00	0,89
0,10		<0,005	0,16	0,54	0,30	<0,005				1,00	1,00	0,83
0,15		0,17	0,58	0,24	0,01					1,00	0,84	0,73
0,20	<0,005	0,43	0,56	0,01						1,00	0,57	0,55
0,25	0,11	0,53	0,36							0,89	0,36	0,36
0,30	0,20	0,56	0,24							0,80	0,24	0,24
0,35	0,28	0,55	0,18							0,72	0,18	0,18
0,40	0,36	0,49	0,15							0,64	0,15	0,15
0,45	0,42	0,45	0,12							0,58	0,12	0,12
0,50	0,47	0,43	0,10							0,53	0,10	0,10
0,60	0,53	0,41	0,06							0,47	0,06	0,06
0,70	0,58	0,38	0,03							0,42	0,03	0,03
0,80	0,64	0,34	0,03							0,37	0,03	0,03
0,90	0,68	0,30	0,02							0,32	0,02	0,02
1,00	0,71	0,27	0,02							0,29	0,02	0,02
Continuous observation along the profile	—											0,91

Table 124

$$b'=0,1 \quad -90° \leqslant \Theta \leqslant 90° \quad d=0,5$$

h in units of d	P_0	P_1	P_2	P_3	P_4	P_5	P_6	P_7	$P_{\geqslant 8}$	$P_{\geqslant 1}$	$P_{\geqslant 2}$	Probability of interception by two profiles
0,05	0,15	<0,005	<0,005	0,01	0,04	0,14	0,22	0,23	0,22	0,85	0,85	0,43
0,10	0,15	0,01	0,12	0,38	0,24	0,04	0,02	0,01	0,03	0,85	0,85	0,40
0,15	0,15	0,11	0,45	0,21	0,03	0,01	0,01	0,01	0,01	0,85	0,74	0,35
0,20	0,15	0,32	0,44	0,05	0,02	0,01	<0,005	<0,005	0,01	0,85	0,53	0,27
0,25	0,19	0,46	0,29	0,03	0,01	0,01	<0,005	<0,005	<0,005	0,81	0,34	0,18
0,30	0,26	0,51	0,20	0,02	0,01	0,01	<0,005	<0,005		0,74	0,24	0,13
0,35	0,32	0,51	0,15	0,02	0,01	0,01	<0,005			0,68	0,17	0,09
0,40	0,38	0,49	0,11	0,01	0,01	<0,005				0,62	0,13	0,07
0,45	0,43	0,47	0,09	0,01	0,01	<0,005				0,57	0,10	0,06
0,50	0,47	0,44	0,07	0,01	<0,005					0,53	0,09	0,05
0,60	0,54	0,40	0,05	0,01	<0,005					0,46	0,06	0,03
0,70	0,60	0,36	0,04	<0,005						0,40	0,04	0,02
0,80	0,64	0,33	0,03	<0,005						0,36	0,03	0,02
0,90	0,68	0,30	0,02	<0,005						0,32	0,02	0,01
1,00	0,70	0,28	0,02							0,30	0,02	0,01
Continuous observation along the profile	0,15											0,44

Table 125

$b'=0,1$ $-30° \leqslant \Theta \leqslant 30°$ $d=0,6$

h in units of d	P_0	P_1	P_2	P_3	P_4	P_5	P_6	P_7	$P_{\geqslant 8}$	$P_{\geqslant 1}$	$P_{\geqslant 2}$	Probability of interception by two profiles
0,05				0,23	0,31	0,32	0,13	<0,005		1,00	1,00	0,57
0,10		0,14	0,56	0,27	0,03					1,00	0,86	0,50
0,15	<0,005	0,56	0,43	0,01						1,00	0,44	0,36
0,20	0,12	0,67	0,21							0,88	0,21	0,21
0,25	0,27	0,60	0,14							0,73	0,14	0,14
0,30	0,36	0,54	0,09							0,64	0,09	0,09
0,35	0,44	0,49	0,07							0,56	0,07	0,07
0,40	0,51	0,44	0,06							0,49	0,06	0,06
0,45	0,56	0,39	0,05							0,44	0,05	0,05
0,50	0,60	0,36	0,04							0,40	0,04	0,04
0,60	0,66	0,32	0,02							0,34	0,02	0,02
0,70	0,70	0,29	0,01							0,30	0,01	0,01
0,80	0,74	0,25	0,01							0,26	0,01	0,01
0,90	0,77	0,22	0,01							0,23	0,01	0,01
1,00	0,79	0,20	0,01							0,21	0,01	0,01
Continuous observation along the profile	—											0,59

Table 126

$b'=0,1$ $-90° \leqslant \Theta \leqslant 90°$ $d=0,6$

h in units of d	P_0	P_1	P_2	P_3	P_4	P_5	P_6	P_7	$P_{\geqslant 8}$	$P_{\geqslant 1}$	$P_{\geqslant 2}$	Probability of interception by two profiles
0,05	0,18	<0,005	0,01	0,11	0,26	0,22	0,11	0,03	0,08	0,82	0,81	0,25
0,10	0,18	0,08	0,41	0,22	0,05	0,02	0,01	0,01	0,02	0,82	0,74	0,22
0,15	0,19	0,37	0,35	0,05	0,02	0,01	0,01	<0,005	0,01	0,81	0,44	0,16
0,20	0,25	0,51	0,19	0,02	0,01	0,01	<0,005	<0,005	<0,005	0,75	0,24	0,09
0,25	0,33	0,52	0,12	0,02	0,01	<0,005	<0,005	<0,005		0,67	0,15	0,06
0,30	0,41	0,48	0,08	0,01	0,01	<0,005	<0,005			0,59	0,10	0,04
0,35	0,48	0,45	0,06	0,01	0,01	<0,005				0,52	0,08	0,03
0,40	0,53	0,41	0,05	0,01	<0,005	<0,005				0,47	0,06	0,02
0,45	0,57	0,38	0,04	0,01	<0,005					0,43	0,05	0,02
0,50	0,61	0,36	0,03	0,01	<0,005					0,39	0,04	0,01
0,60	0,66	0,31	0,02	<0,005						0,34	0,03	0,01
0,70	0,71	0,27	0,02	<0,005						0,29	0,02	0,01
0,80	0,74	0,25	0,01	<0,005						0,26	0,01	0,01
0,90	0,77	0,22	0,01							0,23	0,01	<0,005
1,00	0,79	0,20	0,01							0,21	0,01	<0,005
Continuous observation along the profile	0,18											0,26

Table 127

h in units of d	P_0	P_1	P_2	P_3	P_4	P_5	P_6	P_7	$P_{\geq 8}$	$P_{\geq 1}$	$P_{\geq 2}$	Probability of interception by two profiles
0,05			0,10	0,64	0,22	0,04	<0,005			1,00	1,00	0,34
0,10		0,44	0,52	0,04	<0,005					1,00	0,56	0,25
0,15	0,06	0,80	0,13							0,94	0,13	0,13
0,20	0,27	0,66	0,07							0,73	0,07	0,07
0,25	0,41	0,54	0,05							0,59	0,05	0,05
0,30	0,50	0,47	0,03							0,50	0,03	0,03
0,35	0,57	0,41	0,02							0,43	0,02	0,02
0,40	0,62	0,36	0,02							0,38	0,02	0,02
0,45	0,66	0,32	0,02							0,34	0,02	0,02
0,50	0,69	0,29	0,01							0,31	0,01	0,01
0,60	0,74	0,25	0,01							0,26	0,01	0,01
0,70	0,78	0,22	<0,005							0,22	<0,005	<0,005
0,80	0,80	0,19	<0,005							0,20	<0,005	<0,005
0,90	0,83	0,17	<0,005							0,17	<0,005	<0,005
1,00	0,84	0,15	<0,005							0,16	<0,005	<0,005
Continuous observation along the profile	—											0,36

Table 128

h in units of d	P_0	P_1	P_2	P_3	P_4	P_5	P_6	P_7	$P_{\geq 8}$	$P_{\geq 1}$	$P_{\geq 2}$	Probability of interception by two profiles
0,05	0,22	0,01	0,07	0,34	0,20	0,06	0,03	0,02	0,05	0,78	0,77	0,13
0,10	0,23	0,25	0,38	0,08	0,02	0,01	0,01	0,01	0,01	0,77	0,52	0,10
0,15	0,26	0,52	0,16	0,03	0,01	0,01	<0,005	<0,005	<0,005	0,74	0,22	0,05
0,20	0,37	0,51	0,09	0,02	0,01	<0,005	<0,005	<0,005	<0,005	0,63	0,12	0,03
0,25	0,47	0,46	0,06	0,01	<0,005	<0,005	<0,005			0,53	0,08	0,02
0,30	0,54	0,41	0,04	0,01	<0,005	<0,005				0,46	0,05	0,01
0,35	0,59	0,37	0,03	0,01	<0,005	<0,005				0,41	0,04	0,01
0,40	0,64	0,33	0,02	0,01	<0,005					0,36	0,03	0,01
0,45	0,67	0,31	0,02	<0,005	<0,005					0,33	0,02	0,01
0,50	0,70	0,28	0,02	<0,005						0,30	0,02	<0,005
0,60	0,75	0,24	0,01	<0,005						0,25	0,01	<0,005
0,70	0,78	0,21	0,01	<0,005						0,22	0,01	<0,005
0,80	0,80	0,19	0,01							0,20	0,01	<0,005
0,90	0,83	0,17	<0,005							0,17	<0,005	<0,005
1,00	0,84	0,16	<0,005							0,16	<0,005	<0,005
Continuous observation along the profile	0,22											0,14

Table 129

$$b'=0,1 \qquad -30° \leqslant \Theta \leqslant 30° \qquad d=0,8$$

h in units of d	P_0	P_1	P_2	P_3	P_4	P_5	P_6	P_7	$P_{\geqslant 8}$	$P_{\geqslant 1}$	$P_{\geqslant 2}$	Probability of interception by two profiles
0,05		0,01	— 0,54	0,43	0,01					1,00	0,99	0,16
0,10	0,01	0,74	0,24							0,99	0,24	0,08
0,15	0,22	0,75	0,03							0,78	0,03	0,03
0,20	0,41	0,58	0,02							0,59	0,02	0,02
0,25	0,52	0,46	0,01							0,48	0,01	0,01
0,30	0,60	0,39	0,01							0,40	0,01	0,01
0,35	0,66	0,34	0,01							0,34	0,01	0,01
0,40	0,70	0,30	0,01							0,30	0,01	0,01
0,45	0,73	0,27	<0,005							0,27	<0,005	<0,005
0,50	0,76	0,24	<0,005							0,24	<0,005	<0,005
0,60	0,80	0,20	<0,005							0,20	<0,005	<0,005
0,70	0,83	0,17	<0,005							0,17	<0,005	<0,005
0,80	0,85	0,15	<0,005							0,15	<0,005	<0,005
0,90	0,86	0,13	<0,005							0,14	<0,005	<0,005
1,00	0,88	0,12	<0,005							0,12	<0,005	<0,005
Continuous observation along the profile	—											0,19

Table 130

$$b'=0,1 \qquad -90° \leqslant \Theta \leqslant 90° \qquad d=0,8$$

h in units of d	P_0	P_1	P_2	P_3	P_4	P_5	P_6	P_7	$P_{\geqslant 8}$	$P_{\geqslant 1}$	$P_{\geqslant 2}$	Probability of interception by two profiles
0,05	0,26	0,03	0,26	0,28	0,08	0,03	0,02	0,01	0,03	0,74	0,71	0,06
0,10	0,28	0,40	0,24	0,04	0,02	0,01	0,01	<0,005	0,01	0,72	0,32	0,03
0,15	0,38	0,50	0,08	0,02	0,01	<0,005	<0,005	<0,005	<0,005	0,62	0,12	0,01
0,20	0,49	0,44	0,05	0,01	<0,005	<0,005	<0,005	<0,005		0,51	0,07	0,01
0,25	0,57	0,39	0,03	0,01	<0,005	<0,005	<0,005			0,43	0,04	<0,005
0,30	0,63	0,34	0,02	0,01	<0,005	<0,005				0,37	0,03	<0,005
0,35	0,68	0,30	0,02	<0,005	<0,005					0,32	0,02	<0,005
0,40	0,71	0,27	0,01	<0,005	<0,005					0,29	0,02	<0,005
0,45	0,74	0,25	0,01	<0,005						0,26	0,01	<0,005
0,50	0,77	0,22	0,01	<0,005						0,23	0,01	<0,005
0,60	0,80	0,19	0,01	<0,005						0,20	0,01	<0,005
0,70	0,83	0,17	<0,005							0,17	<0,005	<0,005
0,80	0,85	0,15	<0,005							0,15	<0,005	<0,005
0,90	0,86	0,13	<0,005							0,14	<0,005	<0,005
1,00	0,88	0,12	<0,005							0,12	<0,005	<0,005
Continuous observation along the profile	0,26											0,07

Table 131

$$b'=0{,}1 \qquad -30° \leqslant \Theta \leqslant 30° \qquad d=0{,}9$$

h in units of d	P_0	P_1	P_2	P_3	P_4	P_5	P_6	P_7	$P_{\geqslant 8}$	$P_{\geqslant 1}$	$P_{\geqslant 2}$	Probability of interception by two profiles
0,05	0,01	0,17	0,70	0,13						0,99	0,82	0,04
0,10	0,10	0,83	0,07							0,90	0,07	0,01
0,15	0,36	0,64	<0,005							0,64	<0,005	<0,005
0,20	0,52	0,48	<0,005							0,48	<0,005	<0,005
0,25	0,61	0,39	<0,005							0,39	<0,005	<0,005
0,30	0,68	0,32	<0,005							0,32	<0,005	<0,005
0,35	0,72	0,28	<0,005							0,28	<0,005	<0,005
0,40	0,76	0,24	<0,005							0,24	<0,005	<0,005
0,45	0,78	0,21	<0,005							0,22	<0,005	<0,005
0,50	0,81	0,19	<0,005							0,19	<0,005	<0,005
0,60	0,84	0,16	<0,005							0,16	<0,005	<0,005
0,70	0,86	0,14	<0,005							0,14	<0,005	<0,005
0,80	0,88	0,12	<0,005							0,12	<0,005	<0,005
0,90	0,89	0,11	<0,005							0,11	<0,005	<0,005
1,00	0,90	0,10	<0,005							0,10	<0,005	<0,005
Continuous observation along the profile	<0,005											0,06

Table 132

$$b'=0{,}1 \qquad -90° \leqslant \Theta \leqslant 90° \qquad d=0{,}9$$

h in units of d	$\dot P_0$	P_1	P_2	P_3	P_4	P_5	P_6	P_7	$P_{\geqslant 8}$	$P_{\geqslant 1}$	$P_{\geqslant 2}$	Probability of interception by two profiles
0,05	0,31	0,09	0,33	0,16	0,05	0,02	0,01	0,01	0,02	0,69	0,60	0,01
0,10	0,36	0,45	0,14	0,03	0,01	0,01	<0,005	<0,005	0,01	0,64	0,20	<0,005
0,15	0,48	0,44	0,05	0,01	0,01	<0,005	<0,005	<0,005	<0,005	0,52	0,08	<0,005
0,20	0,58	0,38	0,03	0,01	<0,005	<0,005	<0,005			0,42	0,04	<0,005
0,25	0,65	0,32	0,02	0,01	<0,005	<0,005				0,35	0,03	<0,005
0,30	0,70	0,28	0,01	<0,005	<0,005					0,30	0,02	<0,005
0,35	0,74	0,25	0,01	<0,005	<0,005					0,26	0,01	<0,005
0,40	0,77	0,22	0,01	<0,005						0,23	0,01	<0,005
0,45	0,79	0,20	0,01	<0,005						0,21	0,01	<0,005
0,50	0,81	0,18	0,01	<0,005						0,19	0,01	<0,005
0,60	0,84	0,16	<0,005							0,16	<0,005	<0,005
0,70	0,86	0,14	<0,005							0,14	<0,005	<0,005
0,80	0,88	0,12	<0,005							0,12	<0,005	<0,005
0,90	0,89	0,11	<0,005							0,11	<0,005	<0,005
1,00	0,90	0,10	<0,005							0,10	<0,005	<0,005
Continuous observation along the profile	0,30											0,02

Table 133

b′=0,1   −30° ⩽ Θ ⩽ 30°   d=1,0

h in units of d	P_0	P_1	P_2	P_3	P_4	P_5	P_6	P_7	$P_{\geqslant 8}$	$P_{\geqslant 1}$	$P_{\geqslant 2}$	Probability of interception by two profiles
0,05	0,083	0,287	0,608	0,023						0,918	0,631	
0,10	0,226	0,763	0,011							0,774	0,011	
0,15	0,477	0,523								0,523		
0,20	0,607	0,393								0,393		
0,25	0,686	0,314								0,314		
0,30	0,738	0,262								0,262		
0,35	0,776	0,224								0,224		
0,40	0,804	0,196								0,196		
0,45	0,826	0,174								0,174		
0,50	0,843	0,157								0,157		
0,60	0,869	0,131								0,131		
0,70	0,888	0,112								0,112		
0,80	0,902	0,098								0,098		
0,90	0,913	0,087								0,087		
1,00	0,921	0,079								0,079		
Continuous observation along the profile	0,045											

Table 134

b′=0,1   −90° ⩽ Θ ⩽ 90°   d=1,0

h in units of d	P_0	P_1	P_2	P_3	P_4	P_5	P_6	P_7	$P_{\geqslant 8}$	$P_{\geqslant 1}$	$P_{\geqslant 2}$	Probability of interception by two profiles
0,05	0,372	0,134	0,313	0,099	0,033	0,016	0,009	0,006	0,018	0,628	0,494	
0,10	0,439	0,430	0,091	0,020	0,008	0,004	0,003	0,002	0,003	0,561	0,132	
0,15	0,565	0,380	0,036	0,009	0,004	0,003	0,002	<0,0005		0,435	0,054	
0,20	0,654	0,316	0,021	0,006	0,003	0,001				0,346	0,031	
0,25	0,713	0,267	0,014	0,005	0,001					0,287	0,020	
0,30	0,755	0,231	0,010	0,003	<0,0005					0,245	0,014	
0,35	0,787	0,203	0,008	0,002						0,213	0,010	
0,40	0,811	0,181	0,007	0,001						0,189	0,007	
0,45	0,831	0,164	0,005	<0,0005						0,169	0,005	
0,50	0,847	0,150	0,004							0,153	0,004	
0,60	0,871	0,127	0,002							0,129	0,002	
0,70	0,889	0,111	0,001							0,111	0,001	
0,80	0,902	0,098	<0,0005							0,098	<0,0005	
0,90	0,913	0,087	<0,0005							0,087	<0,0005	
1,00	0,921	0,079								0,079		
Continuous observation along the profile	0,353											

Table 135

$$b'=0,1 \qquad -30° \leqslant \Theta \leqslant 30° \qquad d=1,1$$

h in units of d	P_0	P_1	P_2	P_3	P_4	P_5	P_6	P_7	$P_{\geqslant 8}$	$P_{\geqslant 1}$	$P_{\geqslant 2}$	Probability of interception by two profiles
0,05	0,174	0,356	0,470	0,001						0,826	0,471	
0,10	0,351	0,648	0,001							0,649	0,001	
0,15	0,567	0,433								0,433		
0,20	0,675	0,325								0,325		
0,25	0,740	0,260								0,260		
0,30	0,784	0,216								0,216		
0,35	0,815	0,185								0,185		
0,40	0,838	0,162								0,162		
0,45	0,856	0,144								0,144		
0,50	0,870	0,130								0,130		
0,60	0,892	0,108								0,108		
0,70	0,907	0,093								0,093		
0,80	0,919	0,081								0,081		
0,90	0,928	0,072								0,072		
1,00	0,935	0,065								0,065		
Continuous observation along the profile	0,131											

Table 136

$$b'=0,1 \qquad -90° \leqslant \Theta \leqslant 90° \qquad d=1,1$$

h in units of d	P_0	P_1	P_2	P_3	P_4	P_5	P_6	P_7	$P_{\geqslant 8}$	$P_{\geqslant 1}$	$P_{\geqslant 2}$	Probability of interception by two profiles
0,05	0,432	0,162	0,275	0,069	0,024	0,012	0,007	0,004	0,013	0,568	0,405	
0,10	0,514	0,391	0,065	0,015	0,006	0,003	0,002	0,002	0,002	0,486	0,095	
0,15	0,632	0,327	0,027	0,007	0,003	0,002	0,001	<0,0005		0,367	0,041	
0,20	0,709	0,268	0,016	0,005	0,002	<0,0005				0,291	0,023	
0,25	0,760	0,225	0,011	0,004	0,001					0,240	0,015	
0,30	0,796	0,194	0,008	0,002	<0,0005					0,204	0,010	
0,35	0,822	0,170	0,006	0,001						0,178	0,006	
0,40	0,843	0,152	0,005	<0,0005						0,157	0,005	
0,45	0,859	0,137	0,004	<0,0005						0,141	0,004	
0,50	0,873	0,125	0,002							0,127	0,002	
0,60	0,893	0,106	0,001							0,107	0,001	
0,70	0,908	0,092	<0,0005							0,092	<0,0005	
0,80	0,919	0,081	<0,0005							0,081	<0,0005	
0,90	0,928	0,072	<0,0005							0,072	<0,0005	
1,00	0,935	0,065								0,065		
Continuous observation along the profile	0,412											

Table 137

$$b'=0,1 \qquad -30° \leqslant \Theta \leqslant 30° \qquad d=1,2$$

h in units of d	P_0	P_1	P_2	P_3	P_4	P_5	P_6	P_7	$P_{\geqslant 8}$	$P_{\geqslant 1}$	$P_{\geqslant 2}$	Probability of interception by two profiles
0,05	0,250	0,409	0,341							0,750	0,341	
0,10	0,455	0,545								0,545		
0,15	0,636	0,364								0,364		
0,20	0,727	0,273								0,273		
0,25	0,782	0,218								0,218		
0,30	0,818	0,182								0,182		
0,35	0,844	0,156								0,156		
0,40	0,864	0,136								0,136		
0,45	0,879	0,121								0,121		
0,50	0,891	0,109								0,109		
0,60	0,909	0,091								0,091		
0,70	0,922	0,078								0,078		
0,80	0,932	0,068								0,068		
0,90	0,939	0,061								0,061		
1,00	0,945	0,055								0,055		
Continuous observation along the profile	0,204											

Table 138

$$b'=0,1 \qquad -90° \leqslant \Theta \leqslant 90° \qquad d=1,2$$

h in units of d	P_0	P_1	P_2	P_3	P_4	P_5	P_6	P_7	$P_{\geqslant 8}$	$P_{\geqslant 1}$	$P_{\geqslant 2}$	Probability of interception by two profiles
0,05	0,483	0,187	0,232	0,050	0,019	0,009	0,005	0,003	0,010	0,517	0,330	
0,10	0,577	0,351	0,048	0,012	0,005	0,003	0,002	0,001	0,001	0,423	0,072	
0,15	0,685	0,283	0,021	0,006	0,003	0,002	<0,0005			0,315	0,032	
0,20	0,752	0,230	0,012	0,004	0,002	<0,0005				0,248	0,018	
0,25	0,796	0,192	0,009	0,003	<0,0005					0,204	0,011	
0,30	0,827	0,165	0,007	0,001						0,173	0,008	
0,35	0,850	0,145	0,005	<0,0005						0,150	0,005	
0,40	0,867	0,129	0,004	<0,0005						0,133	0,004	
0,45	0,881	0,117	0,002							0,119	0,002	
0,50	0,892	0,106	0,001							0,108	0,001	
0,60	0,910	0,090	0,001							0,090	0,001	
0,70	0,922	0,078	<0,0005							0,078	<0,0005	
0,80	0,932	0,068	<0,0005							0,068	<0,0005	
0,90	0,939	0,061								0,061		
1,00	0,945	0,055								0,055		
Continuous observation along the profile	0,461											

Table 139

$$b' = 0,1 \qquad -30° \leqslant \Theta \leqslant 30° \qquad d = 1,3$$

h in units of d	P_0	P_1	P_2	P_3	P_4	P_5	P_6	P_7	$P_{\geqslant 8}$	$P_{\geqslant 1}$	$P_{\geqslant 2}$	Probability of interception by two profiles
0,05	0,316	0,439	0,245							0,684	0,245	
0,10	0,535	0,465								0,465		
0,15	0,690	0,310								0,310		
0,20	0,768	0,232								0,232		
0,25	0,814	0,186								0,186		
0,30	0,845	0,155								0,155		
0,35	0,867	0,133								0,133		
0,40	0,884	0,116								0,116		
0,45	0,897	0,103								0,103		
0,50	0,907	0,093								0,093		
0,60	0,923	0,077								0,077		
0,70	0,934	0,066								0,066		
0,80	0,942	0,058								0,058		
0,90	0,948	0,052								0,052		
1,00	0,954	0,046								0,046		
Continuous observation along the profile	0,265											

Table 140

$$b' = 0,1 \qquad -90° \leqslant \Theta \leqslant 90° \qquad d = 1,3$$

h in units of d	P_0	P_1	P_2	P_3	P_4	P_5	P_6	P_7	$P_{\geqslant 8}$	$P_{\geqslant 1}$	$P_{\geqslant 2}$	Probability of interception by two profiles
0,05	0,527	0,205	0,192	0,038	0,015	0,007	0,004	0,003	0,008	0,473	0,268	
0,10	0,630	0,314	0,037	0,009	0,004	0,002	0,002	0,001	<0,0005	0,371	0,056	
0,15	0,728	0,247	0,017	0,005	0,002	0,001	<0,0005			0,272	0,025	
0,20	0,787	0,199	0,010	0,003	0,001					0,213	0,014	
0,25	0,825	0,166	0,007	0,002	<0,0005					0,175	0,009	
0,30	0,852	0,143	0,005	0,001						0,148	0,006	
0,35	0,871	0,125	0,004	<0,0005						0,129	0,004	
0,40	0,886	0,111	0,002							0,114	0,002	
0,45	0,898	0,100	0,002							0,102	0,002	
0,50	0,908	0,091	0,001							0,092	0,001	
0,60	0,923	0,077	<0,0005							0,077	<0,0005	
0,70	0,934	0,066	<0,0005							0,066	<0,0005	
0,80	0,942	0,058								0,058		
0,90	0,948	0,052								0,052		
1,00	0,954	0,046								0,046		
Continuous observation along the profile	0,502											

Table 141

$b'=0,1$ $-30° \leqslant \Theta \leqslant 30°$ $d=1,4$

h in units of d	P_0	P_1	P_2	P_3	P_4	P_5	P_6	P_7	$P_{\geqslant 8}$	$P_{\geqslant 1}$	$P_{\geqslant 2}$	Probability of interception by two profiles
0,05	0,373	0,453	0,174							0,627	0,174	
0,10	0,599	0,401								0,401		
0,15	0,733	0,267								0,267		
0,20	0,800	0,200								0,200		
0,25	0,840	0,160								0,160		
0,30	0,866	0,134								0,134		
0,35	0,886	0,114								0,114		
0,40	0,900	0,100								0,100		
0,45	0,911	0,089								0,089		
0,50	0,920	0,080								0,080		
0,60	0,933	0,067								0,067		
0,70	0,943	0,057								0,057		
0,80	0,950	0,050								0,050		
0,90	0,955	0,045								0,045		
1,00	0,960	0,040								0,040		
Continuous observation along the profile	0,318											

Table 142

$b'=0,1$ $-90° \leqslant \Theta \leqslant 90°$ $d=1,4$

h in units of d	P_0	P_1	P_2	P_3	P_4	P_5	P_6	P_7	$P_{\geqslant 8}$	$P_{\geqslant 1}$	$P_{\geqslant 2}$	Probability of interception by two profiles
0,05	0,565	0,217	0,158	0,030	0,012	0,006	0,003	0,002	0,006	0,435	0,218	
0,10	0,673	0,282	0,030	0,008	0,003	0,002	0,001	0,001	<0,0005	0,327	0,045	
0,15	0,762	0,218	0,013	0,004	0,002	<0,0005				0,238	0,020	
0,20	0,814	0,174	0,008	0,003	<0,0005					0,186	0,011	
0,25	0,848	0,145	0,006	0,001						0,152	0,007	
0,30	0,871	0,124	0,004	<0,0005						0,129	0,005	
0,35	0,888	0,109	0,003	<0,0005						0,112	0,003	
0,40	0,902	0,097	0,002							0,098	0,002	
0,45	0,912	0,087	0,001							0,088	0,001	
0,50	0,920	0,079	0,001							0,080	0,001	
0,60	0,933	0,067	<0,0005							0,067	<0,0005	
0,70	0,943	0,057	<0,0005							0,057	<0,0005	
0,80	0,950	0,050								0,050		
0,90	0,955	0,045								0,045		
1,00	0,960	0,040								0,040		
Continuous observation along the profile	0,538											

Table 143

$$b'=0,1 \qquad -30° \leqslant \Theta \leqslant 30° \qquad d=1,5$$

h in units of d	P_0	P_1	P_2	P_3	P_4	P_5	P_6	P_7	$P_{\geqslant 8}$	$P_{\geqslant 1}$	$P_{\geqslant 2}$	Probability of interception by two profiles
0,05	0,423	0,455	0,121							0,577	0,121	
0,10	0,651	0,349								0,349		
0,15	0,767	0,233								0,233		
0,20	0,825	0,175								0,175		
0,25	0,860	0,140								0,140		
0,30	0,884	0,116								0,116		
0,35	0,900	0,100								0,100		
0,40	0,913	0,087								0,087		
0,45	0,922	0,078								0,078		
0,50	0,930	0,070								0,070		
0,60	0,942	0,058								0,058		
0,70	0,950	0,050								0,050		
0,80	0,956	0,044								0,044		
0,90	0,961	0,039								0,039		
1,00	0,965	0,035								0,035		
Continuous observation along the profile	0,363											

Table 144

$$b'=0,1 \qquad -90° \leqslant \Theta \leqslant 90° \qquad d=1,5$$

h in units of d	P_0	P_1	P_2	P_3	P_4	P_5	P_6	P_7	$P_{\geqslant 8}$	$P_{\geqslant 1}$	$P_{\geqslant 2}$	Probability of interception by two profiles
0,05	0,598	0,225	0,129	0,024	0,009	0,005	0,003	0,002	0,005	0,402	0,177	
0,10	0,710	0,254	0,024	0,006	0,003	0,002	0,001	<0,0005		0,290	0,036	
0,15	0,791	0,193	0,011	0,003	0,002	<0,0005				0,209	0,016	
0,20	0,837	0,154	0,007	0,002	<0,0005					0,163	0,009	
0,25	0,867	0,128	0,005	0,001						0,133	0,006	
0,30	0,887	0,109	0,003	<0,0005						0,113	0,004	
0,35	0,902	0,096	0,002							0,098	0,002	
0,40	0,914	0,085	0,001							0,086	0,001	
0,45	0,923	0,076	0,001							0,077	0,001	
0,50	0,931	0,069	<0,0005							0,069	<0,0005	
0,60	0,942	0,058	<0,0005							0,058	<0,0005	
0,70	0,950	0,050								0,050		
0,80	0,956	0,044								0,044		
0,90	0,961	0,039								0,039		
1,00	0,965	0,035								0,035		
Continuous observation along the profile	0,569											

Table 145

$$b'=0,1 \qquad -30° \leqslant \Theta \leqslant 30° \qquad d=1,7$$

h in units of d	P_0	P_1	P_2	P_3	P_4	P_5	P_6	P_7	$P_{\geqslant 8}$	$P_{\geqslant 1}$	$P_{\geqslant 2}$	Probability of interception by two profiles
0,05	0,509	0,439	0,052							0,491	0,052	
0,10	0,728	0,272								0,272		
0,15	0,819	0,181								0,181		
0,20	0,864	0,136								0,136		
0,25	0,891	0,109								0,109		
0,30	0,909	0,091								0,091		
0,35	0,922	0,078								0,078		
0,40	0,932	0,068								0,068		
0,45	0,940	0,060								0,060		
0,50	0,946	0,054								0,054		
0,60	0,955	0,045								0,045		
0,70	0,961	0,039								0,039		
0,80	0,966	0,034								0,034		
0,90	0,970	0,030								0,030		
1,00	0,973	0,027								0,027		
Continuous observation along the profile	0,438											

Table 146

$$b'=0,1 \qquad -90° \leqslant \Theta \leqslant 90° \qquad d=1,7$$

h in units of d	P_0	P_1	P_2	P_3	P_4	P_5	P_6	P_7	$P_{\geqslant 8}$	$P_{\geqslant 1}$	$P_{\geqslant 2}$	Probability of interception by two profiles
0,05	0,653	0,229	0,085	0,016	0,007	0,003	0,002	0,001	0,003	0,347	0,118	
0,10	0,768	0,207	0,016	0,005	0,002	0,001	<0,0005			0,232	0,025	
0,15	0,834	0,155	0,008	0,003	0,001					0,166	0,011	
0,20	0,871	0,122	0,005	0,001						0,129	0,006	
0,25	0,895	0,101	0,003	<0,0005						0,105	0,004	
0,30	0,911	0,087	0,002							0,089	0,002	
0,35	0,923	0,075	0,001							0,077	0,001	
0,40	0,933	0,067	0,001							0,067	0,001	
0,45	0,940	0,060	<0,0005							0,060	<0,0005	
0,50	0,946	0,054	<0,0005							0,054	<0,0005	
0,60	0,955	0,045								0,045		
0,70	0,961	0,039								0,039		
0,80	0,966	0,034								0,034		
0,90	0,970	0,030								0,030		
1,00	0,973	0,027								0,027		
Continuous observation along the profile	0,620											

Table 147

$b'=0,1$ $\quad$ $-30° \leqslant \Theta \leqslant 30°$ $\quad$ $d=2,0$

h in units of d	P_0	P_1	P_2	P_3	P_4	P_5	P_6	P_7	$P_{\geqslant 8}$	$P_{\geqslant 1}$	$P_{\geqslant 2}$	Probability of interception by two profiles
0,05	0,613	0,381	0,006							0,387	0,006	
0,10	0,804	0,196								0,196		
0,15	0,869	0,131								0,131		
0,20	0,902	0,098								0,098		
0,25	0,921	0,079								0,079		
0,30	0,935	0,065								0,065		
0,35	0,944	0,056								0,056		
0,40	0,951	0,049								0,049		
0,45	0,956	0,044								0,044		
0,50	0,961	0,039								0,039		
0,60	0,967	0,033								0,033		
0,70	0,972	0,028								0,028		
0,80	0,975	0,025								0,025		
0,90	0,978	0,022								0,022		
1,00	0,980	0,020								0,020		
Continuous observation along the profile	0,522											

Table 148

$b'=0,1$ $\quad$ $-90° \leqslant \Theta \leqslant 90°$ $\quad$ $d=2,0$

h in units of d	P_0	P_1	P_2	P_3	P_4	P_5	P_6	P_7	$P_{\geqslant 8}$	$P_{\geqslant 1}$	$P_{\geqslant 2}$	Probability of interception by two profiles
0,05	0,719	0,215	0,045	0,010	0,004	0,002	0,001	0,001	0,002	0,281	0,066	
0,10	0,827	0,158	0,010	0,003	0,002	0,001	<0,0005			0,173	0,015	
0,15	0,878	0,115	0,005	0,002	<0,0005					0,122	0,007	
0,20	0,906	0,091	0,003	<0,0005						0,094	0.004	
0,25	0,923	0,075	0,002							0,077	0,002	
0,30	0,935	0,064	0,001							0,065	0,001	
0,35	0,944	0,055	<0,0005							0,056	<0,0005	
0,40	0,951	0,049	<0,0005							0,049	<0,0005	
0,45	0,956	0,044	<0,0005							0,044	<0,0005	
0,50	0,961	0,039								0,039		
0,60	0,967	0,033								0,033		
0,70	0,972	0,028								0,028		
0,80	0,975	0,025								0,025		
0,90	0,978	0,022								0,022		
1,00	0,980	0,020								0,020		
Continuous observation along the profile	0,677											

NOMOGRAMS

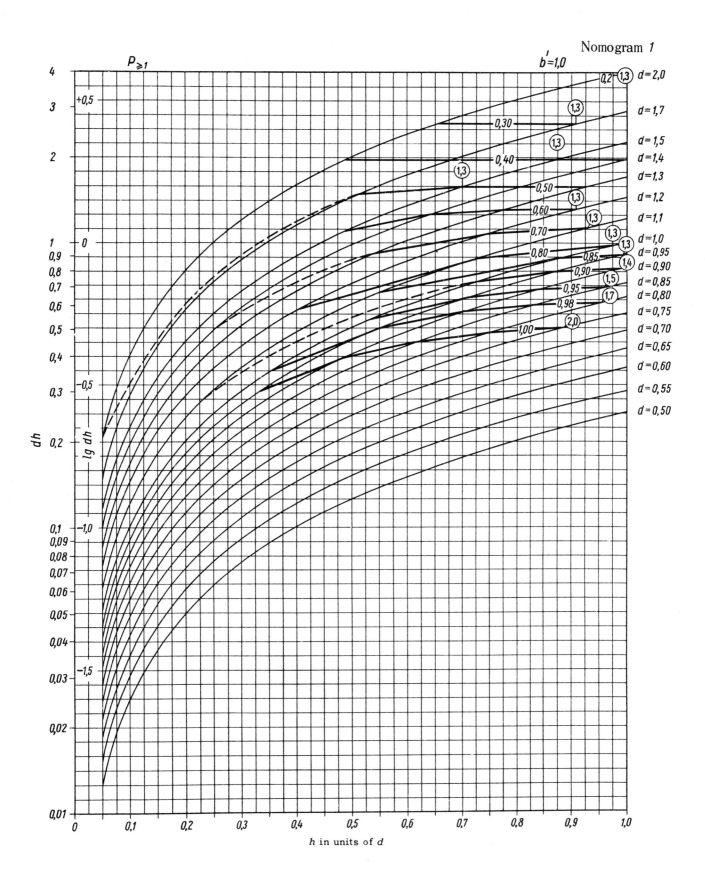

Nomogram *1*

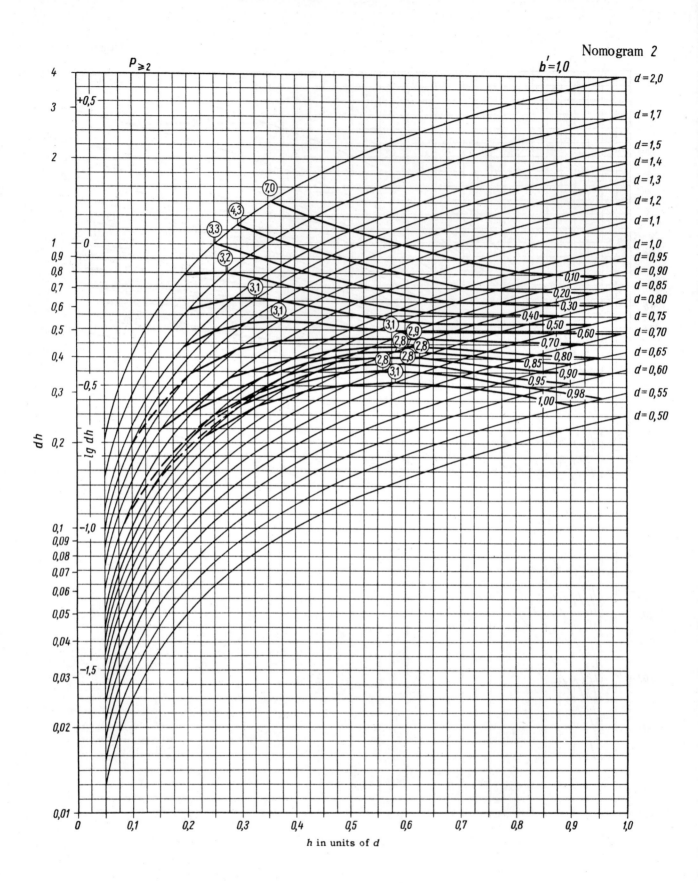

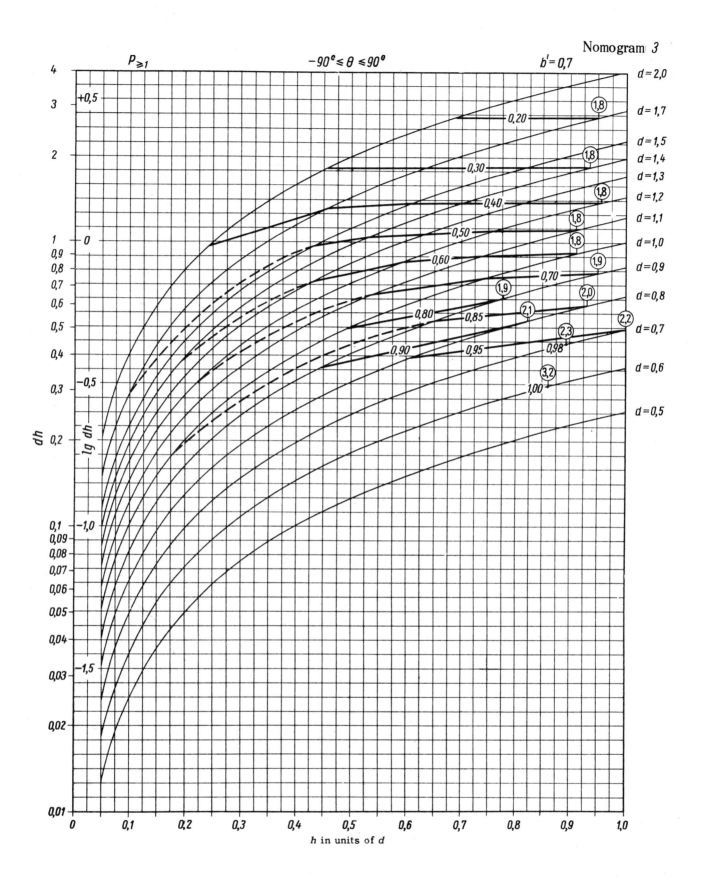

$P_{\geq 1}$ $-90° \leq \theta \leq 90°$ $b' = 0,7$

dh

$lg\ dh$

h in units of d

d = 2,0
d = 1,7
d = 1,5
d = 1,4
d = 1,3
d = 1,2
d = 1,1
d = 1,0
d = 0,9
d = 0,8
d = 0,7
d = 0,6
d = 0,5

0,20
0,30
0,40
0,50
0,60
0,70
0,80
0,85
0,90
0,95
0,98
1,00

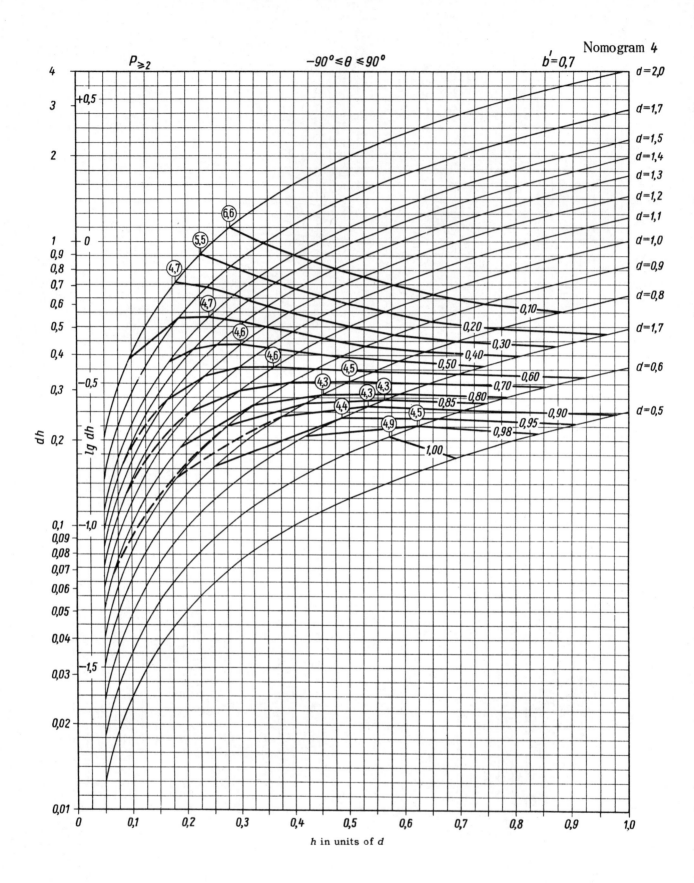

$P_{\geqslant 2}$ $-90° \leqq \theta \leqq 90°$ $b' = 0,7$

h in units of d

92

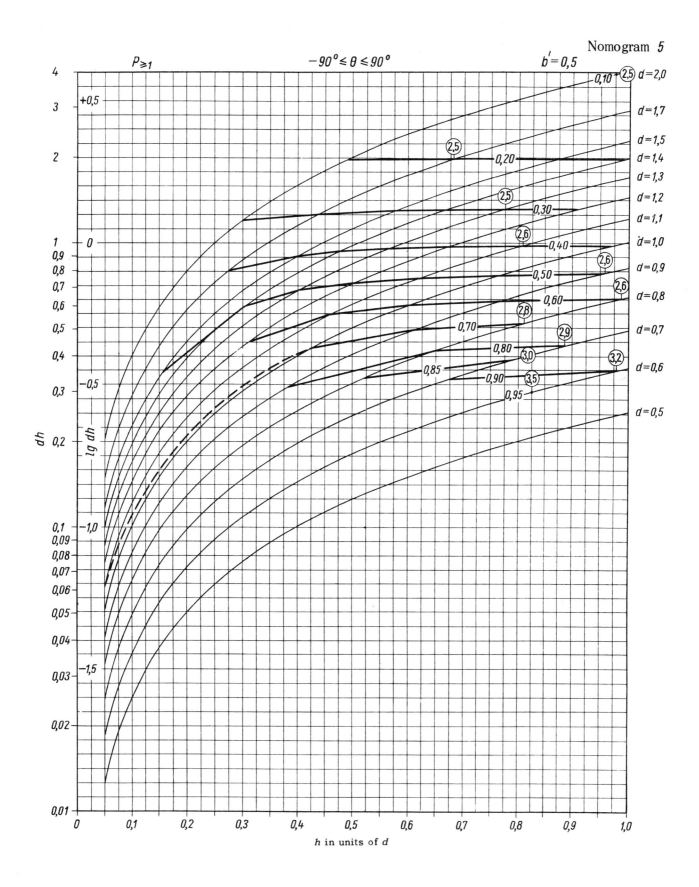

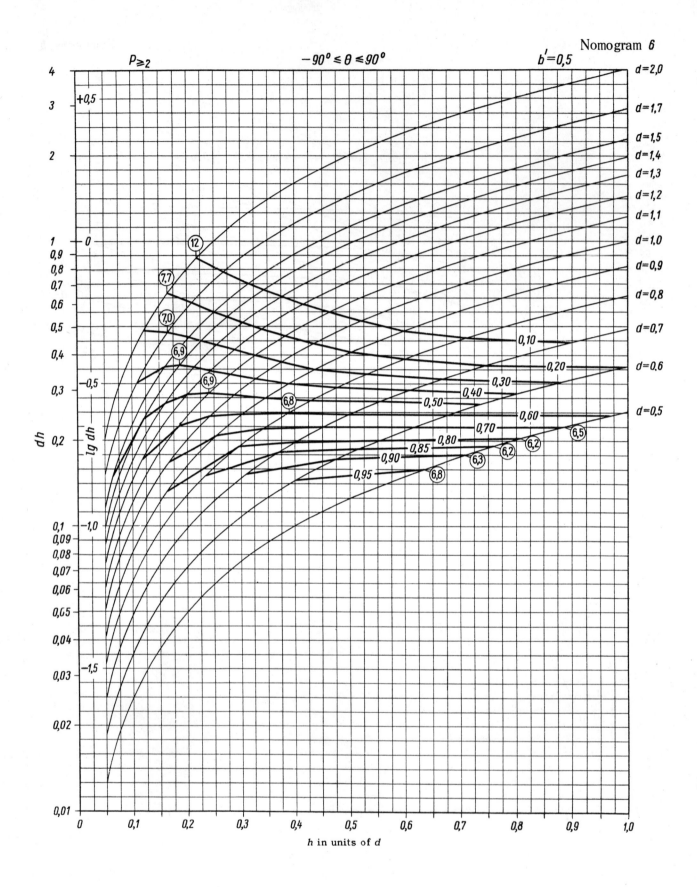

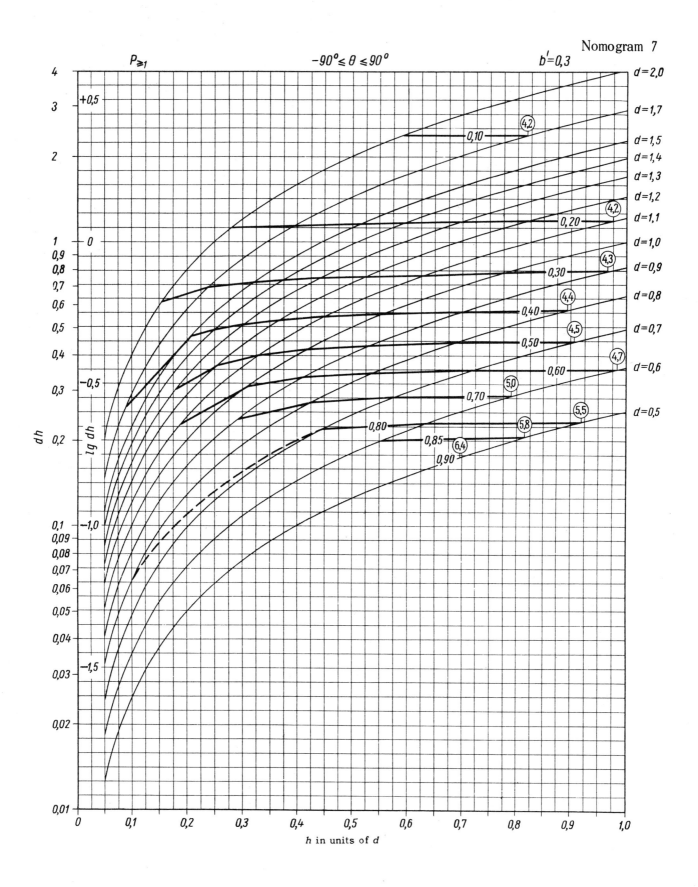

Nomogram 7

$P_{\geqq 1}$

$-90^\circ \leqq \theta \leqq 90^\circ$

$b' = 0,3$

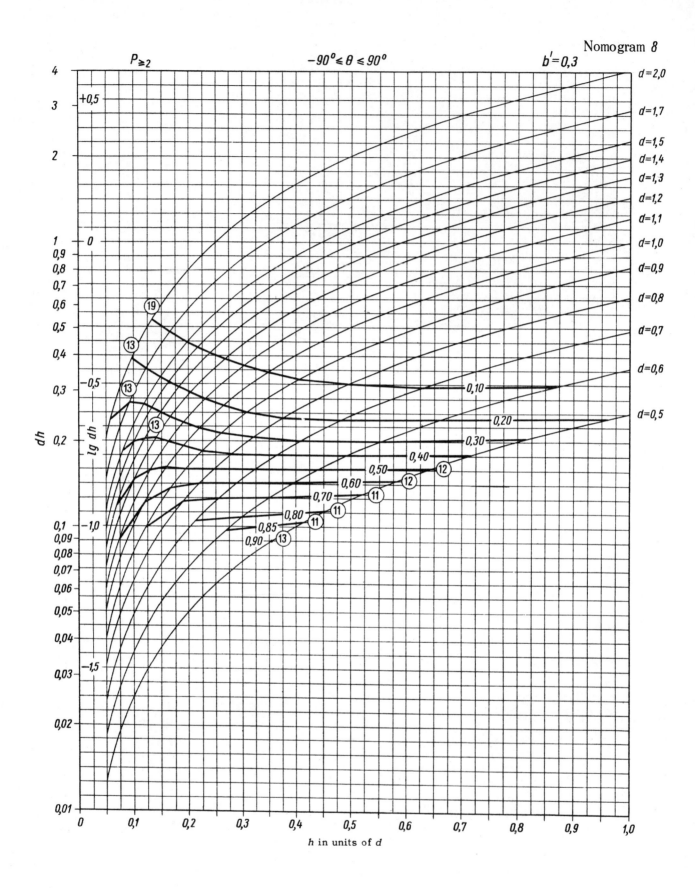

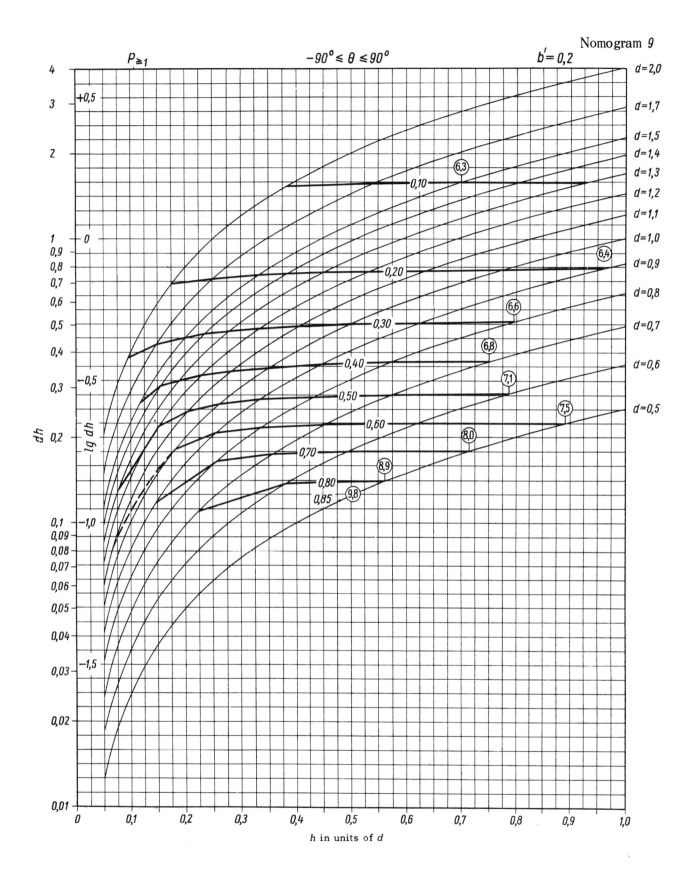

$P_{\geq 1}$ $-90° \leq \theta \leq 90°$ $b' = 0,2$

h in units of *d*

97

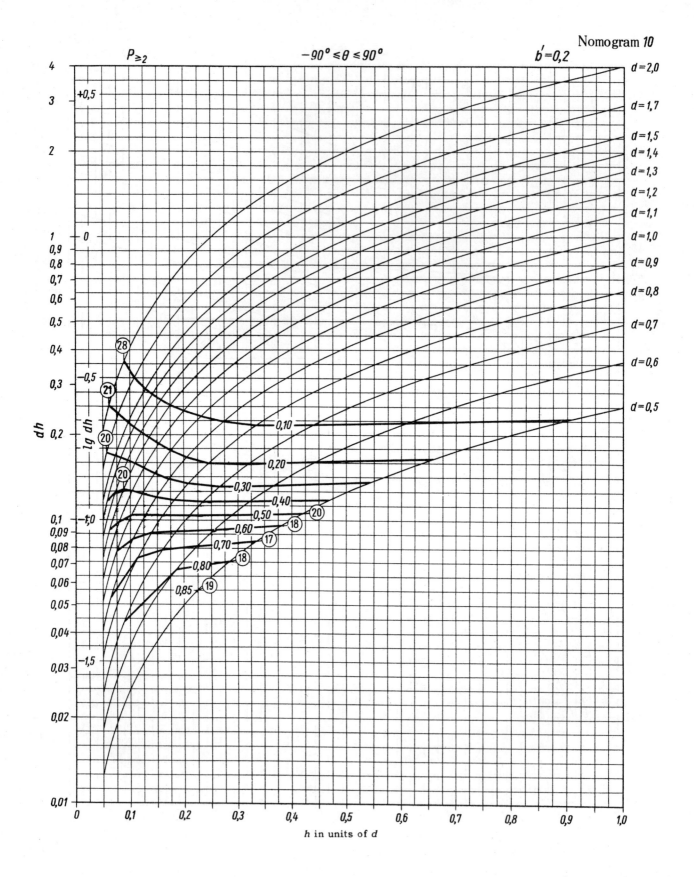

Nomogram *10*

$P_{\geqslant 2}$

$-90° \leqslant \theta \leqslant 90°$

$b' = 0,2$

dh

lg dh

h in units of *d*

98

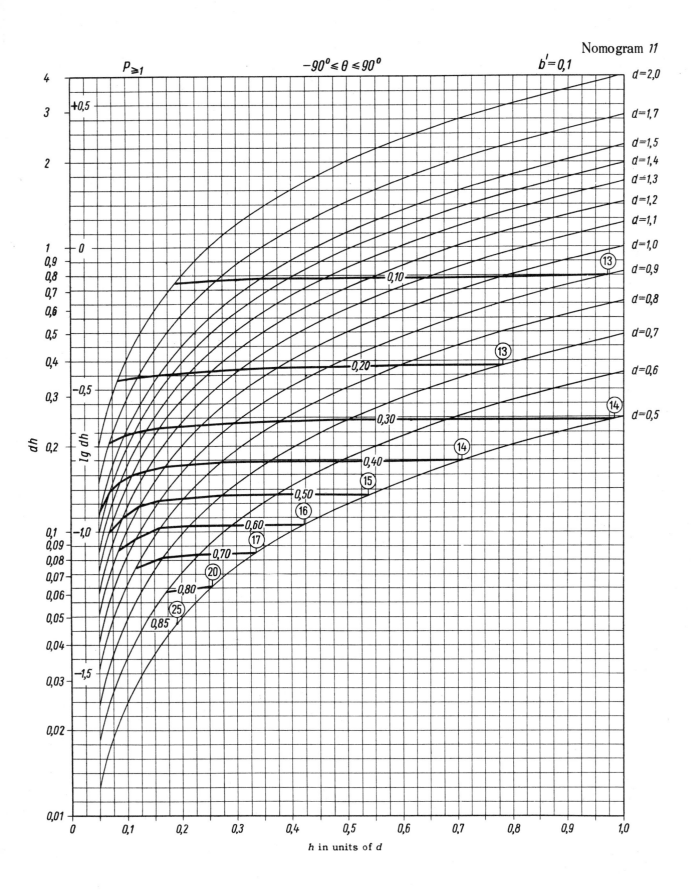

$P_{\geqslant 1}$ $\qquad -90^{\circ} \leqslant \theta \leqslant 90^{\circ} \qquad$ $b' = 0,1$

h in units of d

Univ. of Arizona Library

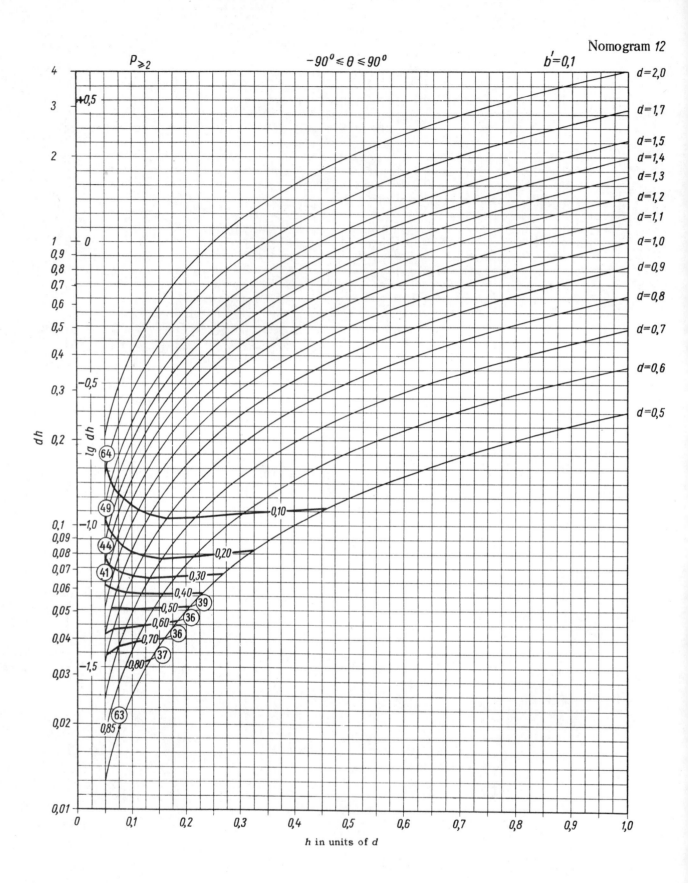

$P_{\geqslant 2}$ $-90^o \leqslant \theta \leqslant 90^o$ $b'=0,1$

$d=2,0$

$d=1,7$

$d=1,5$

$d=1,4$

$d=1,3$

$d=1,2$

$d=1,1$

$d=1,0$

$d=0,9$

$d=0,8$

$d=0,7$

$d=0,6$

$d=0,5$

dh

$lg\ dh$

h in units of d

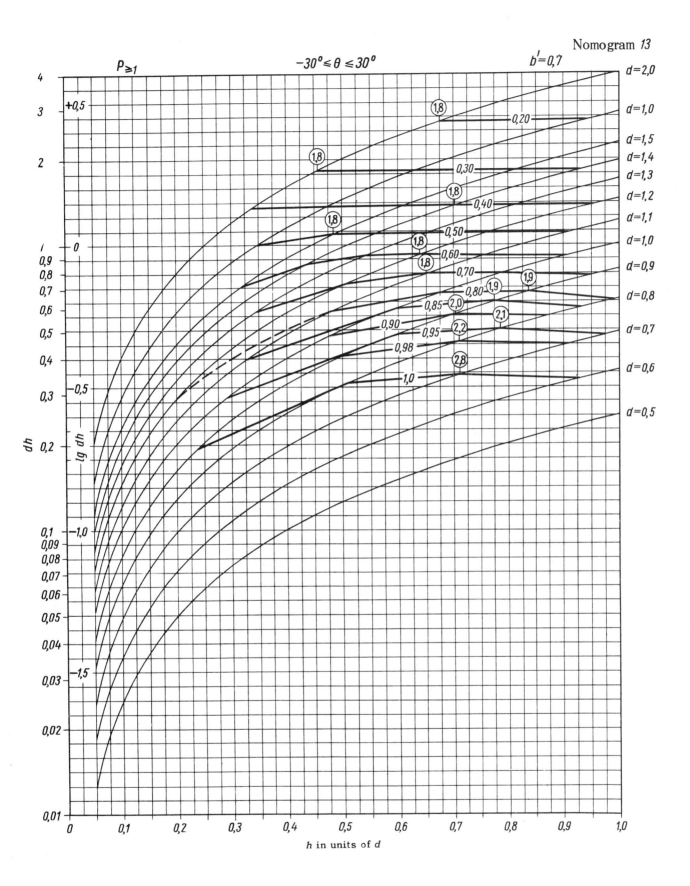

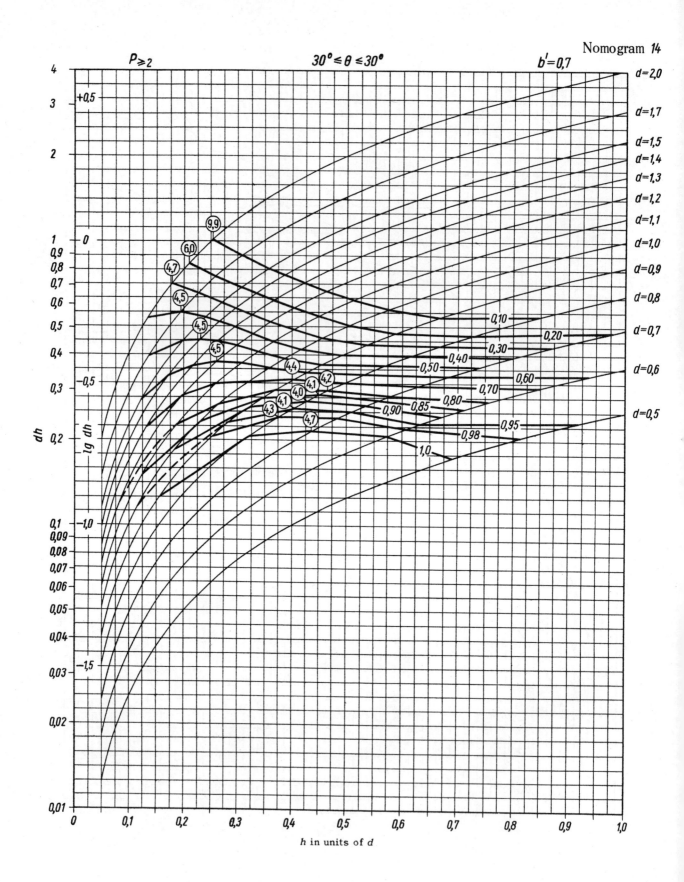

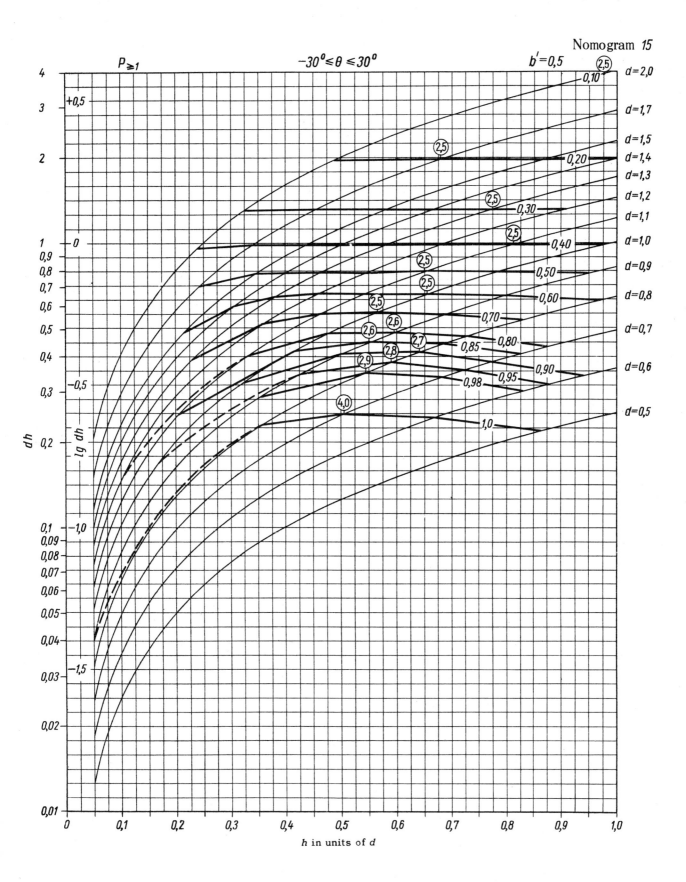

$P_{\geqslant 1}$

$-30^{\circ} \leqslant \theta \leqslant 30^{\circ}$

$b'=0,5$

h in units of d

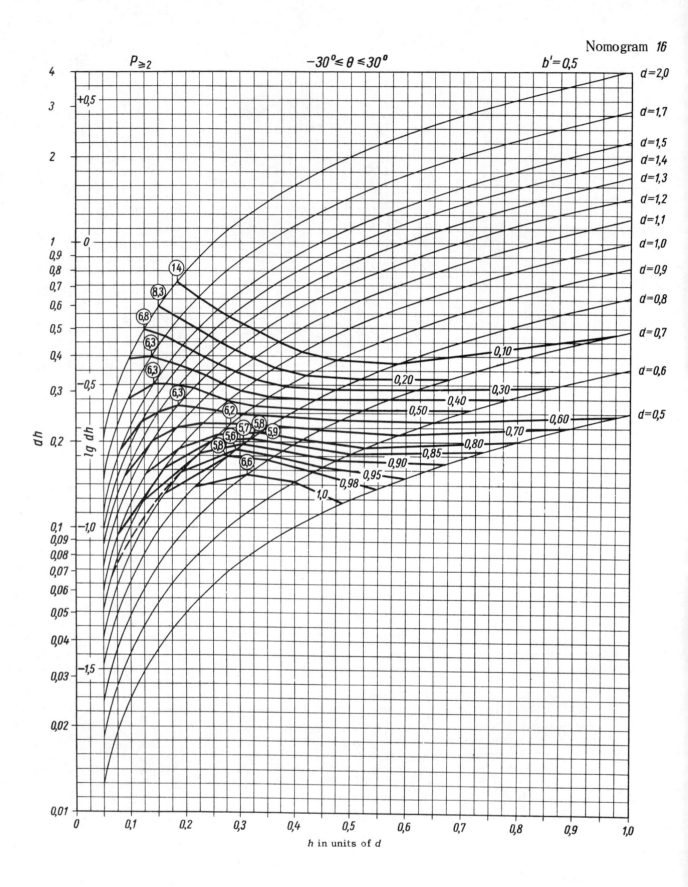

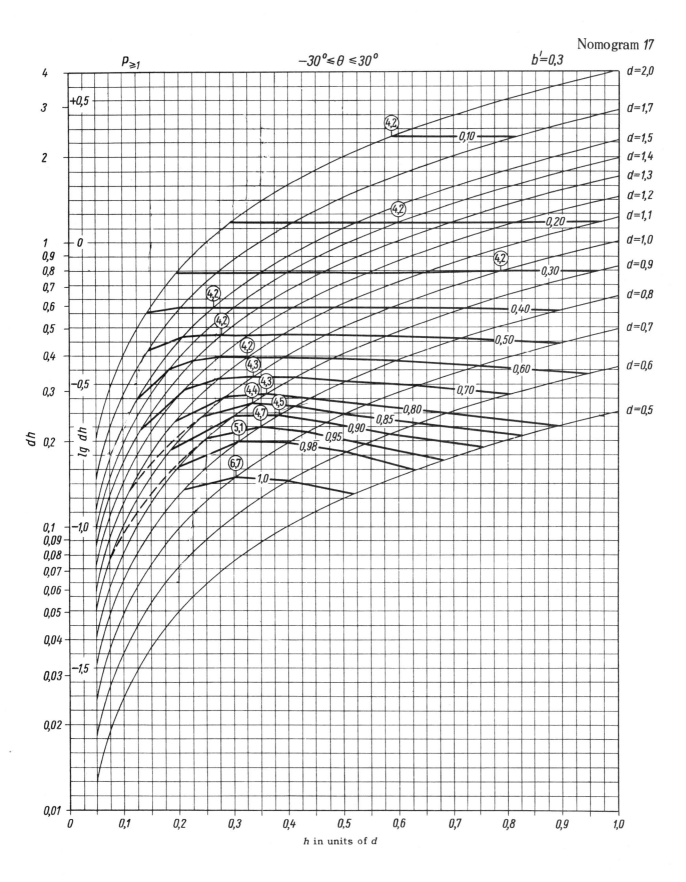

$P_{\geqslant 1}$ $\quad -30° \leq \theta \leq 30°$ $\quad b'=0,3$

$d=2,0$
$d=1,7$
$d=1,5$
$d=1,4$
$d=1,3$
$d=1,2$
$d=1,1$
$d=1,0$
$d=0,9$
$d=0,8$
$d=0,7$
$d=0,6$
$d=0,5$

dh

h in units of d

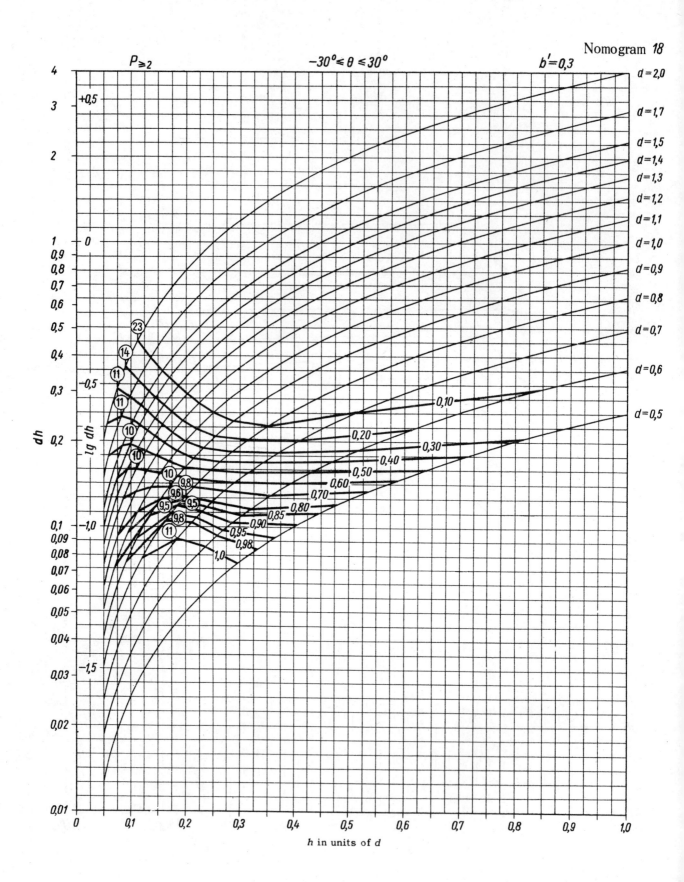

$P_{\geqq 2}$ $-30° \leqslant \theta \leqslant 30°$ $b'=0,3$

dh

$lg\ dh$

h in units of d

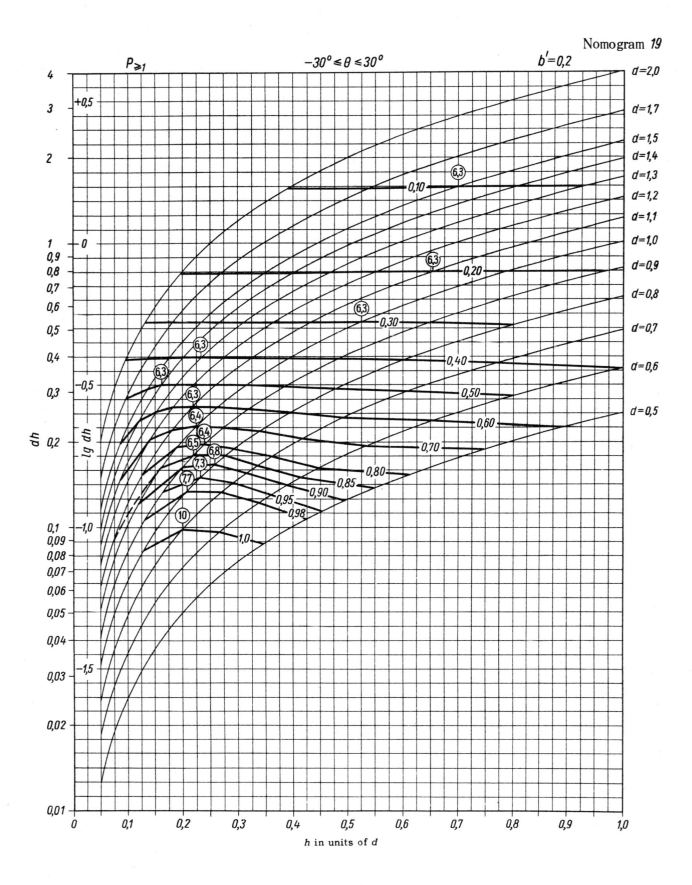

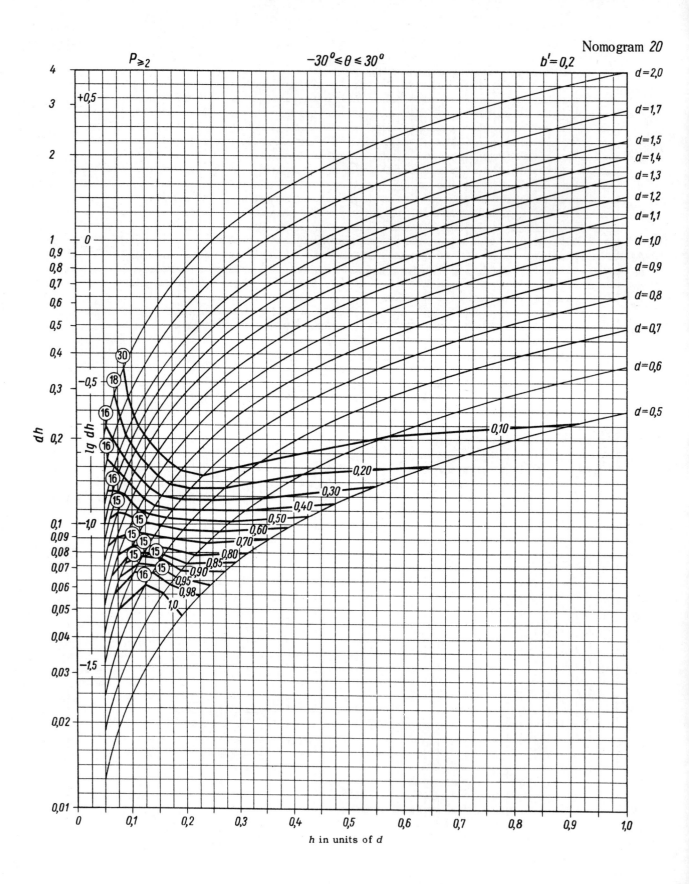

$P_{\geqslant 2}$ $-30^\circ \leqslant \theta \leqslant 30^\circ$ $b' = 0,2$

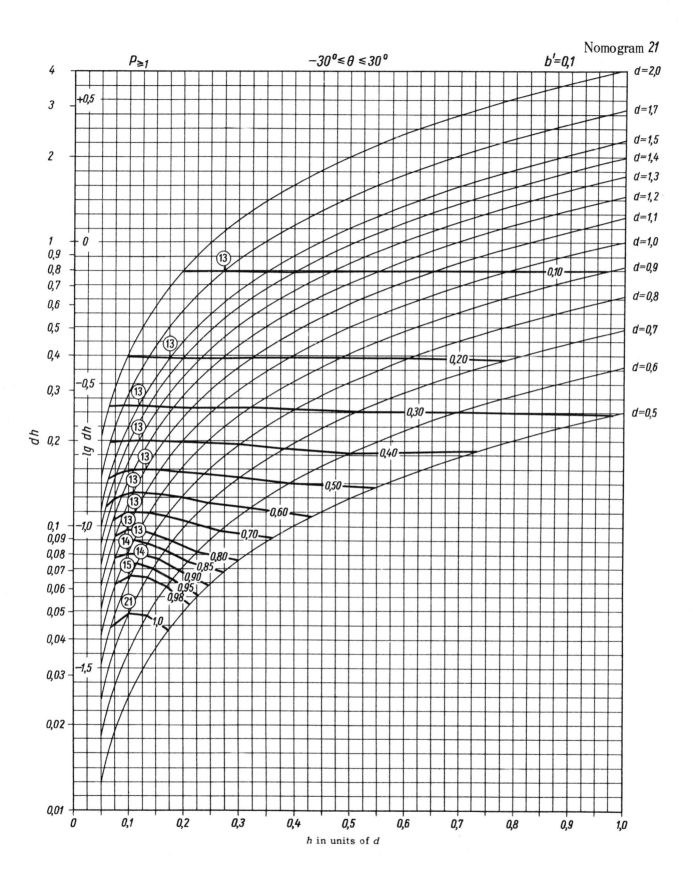

$P_{\geqslant 1}$ $-30° \leqslant \theta \leqslant 30°$ $b'=0,1$

h in units of *d*

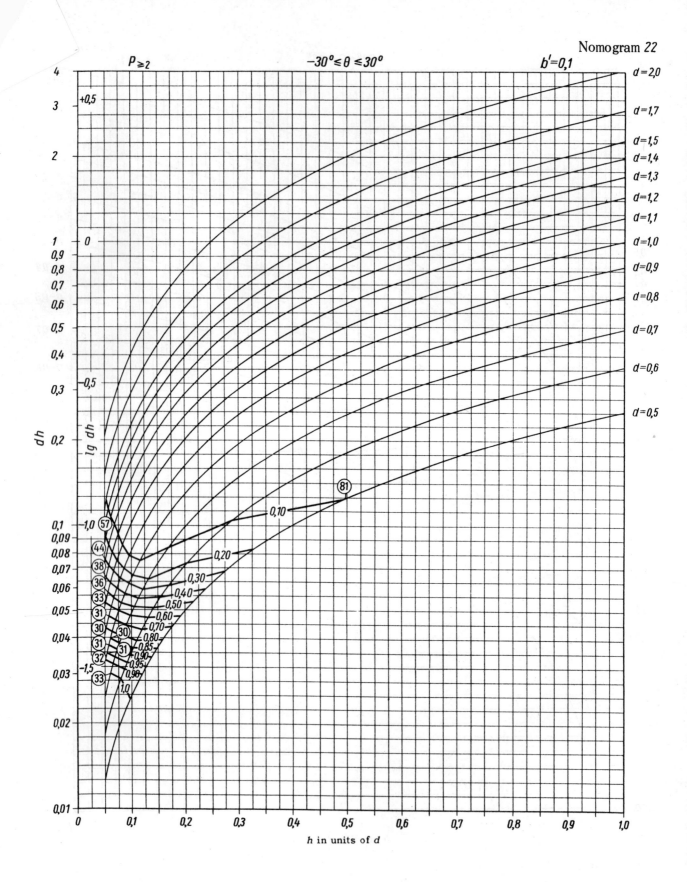

$P_{\geq 2}$ $-30° \leq \theta \leq 30°$ $b'=0,1$

dh

$lg\ dh$

h in units of d